American
Politics
and the
Environment

American Politics

Politics

and the

Environment

SECOND EDITION

Byron W. Daynes, Glen Sussman,
and Jonathan P. West

First edition published 2002.

Published by State University of New York Press, Albany

© 2016 State University of New York

For information, contact State University of New York Press, Albany, NY
www.sunypress.edu

Production, Ryan Morris
Marketing, Michael Campochiaro

Library of Congress Cataloging-in-Publication Data

Daynes, Byron W.
    American politics and the environment / Byron W. Daynes, Glen Sussman, and Jonathan P. West. — Second edition.
        pages cm
    Includes bibliographical references and index.
    ISBN 978-1-4384-5933-2 (hc : alk. paper)—978-1-4384-5932-5 (pb : alk. paper)
    ISBN 978-1-4384-5934-9 (e-book)
    1. Environmental policy—United States.   I. Sussman, Glen.   II. West, Jonathan P. (Jonathan Page), 1941–   III. Title.

GE180.S87 2016
363.7'0560973—dc23   2015008205

10 9 8 7 6 5 4 3 2 1

*To Byron "Bill" Daynes, a beloved friend,
valued colleague, and respected scholar*

Contents

List of Figures and Tables ix

Chapter 1
The American Political Setting and the Environment 1

Chapter 2
American Federalism and Environmental Politics 21
    Personal Profile: Mary Nichols and CARB: Environmental
        Champions 28
    Case Study: The Dynamics of Implementation and
        Enforcement I: Mountaintop Removal in West Virginia 33
    Case Study: The Dynamics of Implementation and
        Enforcement II: Climate Change and Sea-Level Rise in
        South Florida 39

Chapter 3
Public Opinion, Interest Groups, and the Environment 47
    Case Study: Climate Change and Public Opinion 50
    Personal Profile: Robert F Kennedy Jr. 56
    Case Study: Interest Group Activism and the Place of
        Wolves in the Rocky Mountain Regions 65

Chapter 4
Congress, the Legislative Process, and the Environment 71
    Personal Profile: Senator Susan Collins: Environmental Defender 81
    Case Study: Cap and Trade: Failure to Pass a Climate Bill 85

Chapter 5
The Environmental Presidency   99
    Personal Profile: FDR and the "Golden Age"   99
    Case Study: Leaving a Legacy: Barack Obama and
        Climate Change   104

Chapter 6
Executive Agencies and the Environment   123
    Personal Profile: Steven Chu: The Scientist in Charge   127
    Case Study: EPA Regulations of Carbon Emissions   131

Chapter 7
The Environmental Court   149
    Personal Profile: The Environment: Friend and Supporter
        versus a "Slash-and-Burn" Opponent   159
    Case Study: The Court and "Those Dam Fish":
        TVA v. Hill (1978)   164
    Case Study: Climate Crisis is Real, Mr. Limbaugh, REAL!   166

Chapter 8
The Global Environment   171
    Guest Essay: Solidarity Norms and International Climate
        Change Cooperation, by Christina Slentz   182
    Case Study: Biodiversity and Endangered Species   199

Chapter 9
American Politics and the Environment: Conclusion   205

Notes   225

Index   249

Figures and Tables

Figure

Figure 5.1 Presidential Roles   102

Tables

Table 3.1 Support for Conventional and Alternative Sources
           of Energy by Party   53

Table 3.2 Campaign Contributions to Congressional Democrats
           and Republicans by Environmental Interest Groups,
           1994–2014   62

Table 5.1 Number of International Agreements (by President)
           Dealing with the Environment as a Percentage of all
           Agreements, 1949–1996   118

Table 5.2 Presidential Types Based on their Approach to
           Environmental Policy   120

Table 6.1 Strengths and Challenges Reported by EPA Agency
           Employees (Average percent)   142

Table 6.2 Federal Government and EPA Workplace Satisfaction
           (Average percent)   143

Table 8.1 Citizens' Views of the Importance of the Environment
           (Selected Countries)   174

Table 8.2 Summary Chart (Gray shading indicates above sample
           average levels)   187

Table 8.3   Group THREE (Gray shading indicates above sample average levels)   190

Table 8.4   The North American States (Gray shading indicates above sampling average levels)   193

1

The American Political Setting and the Environment

Environmentalism is one among many complex and technical policy issues that has challenged political leaders and citizens alike since the dawn of the Industrial Revolution. As one journalist specializing on U.S. environmental policy observed, "The economic prosperity of the Industrial Revolution—indeed the rise of America—came at a steep price: lost wilderness, contaminated waters, dirty skies, endangered animals and plants."[1] By the mid-1960s, the modern American environmental movement focused not only on domestic concerns but also included transnational environmental policy issues ranging from acid rain to stratospheric ozone depletion to global warming and climate change. In short, it became increasingly clear that the United States and other countries were exponentially threatening the health of the environment at home and abroad.

To what extent have U.S. public officials included environmental issues as a central feature of the public agenda? For some, the question of environmental protection concerns value conflicts between preservation and development, where tradeoffs are demanded of contending forces. While some public officials have advocated that the federal government play a strong role in protecting the environment, a limited number of their colleagues are reluctant to impose governmental authority over business and industry with respect to the environment. Still others argue that state and local governments rather than the federal government should play the primary role in managing environmental affairs.

The history of the environmental policy process in the United States has been associated with state-level politics where policymakers, more often than not, have supported economic development over environmental protection. Over the last half-century, however, the federal government assumed increasing responsibility for managing environmental affairs. At the same time, public opinion informs us that American citizens have supported protecting the environment over economic development.[2] Moreover, Americans are more likely to prefer that the federal government take action to protect the environment, rather than rely on business and industry to do so.[3]

The political struggle regarding the environment is framed within the American constitutional system of government involving the three major institutions of the federal government. A secondary consideration involves federalism and the extent to which the national government and the fifty state governments should play a role in environmental management. The environment as an important public policy issue also includes the debate over the extent to which science should be involved in environmental policymaking. Consequently, environmental management can be viewed as being subject to a variety of influences that have affected the decision-making process.

The American Political System, Public Policymaking, and the Environment

In the American political system, public policy is subject to a variety of political constraints including but not limited to the dispersion of power prescribed by the Madisonian model of separation of power and the system of checks and balances. The federal system of government divides political power between the national government and the fifty states. Moreover, as the framers of the Constitution were well aware when they argued in *Federalist #10*, the governmental system was subject to pressures exerted by organized interests. This motivated the framers to design a system that would moderate the actions of the myriad political actors within the system.

In the American political setting, the three major national institutions (legislative, executive, judicial) have specific areas of political responsibility yet also exert their influence beyond their respective jurisdictions. Congress has the power to pass legislation, yet the framers of the Constitution gave to the president the ability to negate the efforts of those 535 legislators though the power of the veto. Then again, Congress can override the president's veto power if it can muster sufficient support (two-thirds of the congressional membership) to oppose the president's actions. Furthermore, the Supreme Court can exercise its power of judicial review in response to actions taken by the other two institutions.

Congress is a decentralized institution in which political power is fragmented among a variety of committees and subcommittees that can promote, delay, or oppose legislation as well as expand their jurisdiction. For example, several different committees and subcommittees in the House and the Senate have jurisdiction over environmental affairs. Consequently, notwithstanding congressional responsibility for advancing the national interest, members of the legislative branch of government remain committed to protecting state and local interests. In the process of doing so, important issues at the national level may become subverted by subnational pressure. In addition to these considerations, Congress is also influenced by the partisan makeup of the legislative body. Although bipartisanship is evident on some legislation, partisan conflict over public policy is an integral feature of the legislative process. As far as Congress and the environment are concerned, the "golden age" of environmental legislation occurred during the 1960s and 1970s. During this period, as a result of bipartisanship among legislators, important bills (some modest, others substantive) were passed by Congress and signed into law. This legislation, some with subsequent amendments added, included the Clean Air Act (1963, 1970, 1977), Wilderness Act (1964), Endangered Species Conservation Act (1966, 1973), National Environmental Policy Act (1970), Marine Mammal Protection Act (1972), Clean Water Act (1972, 1977), Safe Drinking Water Act (1974, 1986), and the Superfund (1980, 1986).

Although certain presidents have used the power resources available to them to take action on behalf of the environment—signing legislation, issuing executive orders, using the veto power—the environment has yet to assume a central place in their legislative agenda. Later in this book, we will discuss the classification of presidents as *activist* or *symbolic* in their behavior toward the environment. As an activist, the president can take actions that promote environmental protection or support a developmental ethic over conservation efforts. Or, the president can respond to environmental challenges in a symbolic way, exhibiting only modest to little attention to environmental challenges.

Moreover, just as the president sits atop the executive branch of government and sets the public agenda, executive agencies also play an important role in the policymaking process. The bureaucracy is similar to the legislative branch, in that it is a decentralized institution comprised of numerous agencies, departments, and bureaus sometimes having overlapping jurisdiction. As a public policy area, the environment is under the jurisdiction of a variety of regulatory agencies that either cooperate or engage in turf wars.[4]

As well, executive agencies, including major players such as the Environmental Protection Agency (EPA) and the Department of the Interior, have been politicized as a result of presidential budget priorities and the appointment process. During the 1980s, for instance, both offices received considerable news media attention due to problems arising over political and personal matters. Anne Gorsuch, EPA administrator, resigned and Rita Lavelle, assistant administrator for hazardous waste was fired, due to allegations of mismanagement and lax enforcement of environmental regulations.[5] Secretary of the Interior James Watt had what Robert Durant called a "confrontational, arrogant, and badgering style" that "fanned the flames of conflict with congressional oversight committees . . . and the environmental community."[6] Although Watt eventually resigned in response to mounting pressure from environmentalists, citizens, and members of Congress, Watt's protégé Gale Norton was later nominated by President George W. Bush to serve as his Secretary of the Interior. On the other hand, Bush's initial appointment of Christine Todd Whitman to head the EPA was viewed positively since she had a background of being sympathetic to environmental concerns. However, her tenure was relatively short since her views were increasingly at odds with the administration. As Kristina Horn Sheeler informed us, when Whitman accepted the position of EPA administrator, she made reference to Teddy Roosevelt, "our first conservationist president," who "understood the necessity of striking the right balance between competing interests for the good of all Americans."[7] The notion of Bush employing a balanced approach was quickly forgotten as Whitman was characterized as "shoved to the margins," "undercut," "undermined," "isolated," "the odd woman out," "out of step," and the "lone voice."[8] In contrast, President Barack Obama appointed Lisa Jackson to the head the EPA. Jackson was a strong, committed EPA administrator who used the regulatory process in support of ecological issues especially in the policy domain of climate change and its threat to the environment and to public health.

While Congress, the presidency, and executive agencies are characterized as political institutions, the Supreme Court exercises its authority through the judicial process. The Court, third pillar of the country's national institutional framework, has an important role in influencing the actions of the other two branches as well as the states. As a result of the 1803 *Marbury v. Madison* decision, the Court has the power of judicial review, which underlies its ability to interpret the actions of the executive and legislative branches of government as well as events at the subnational level. As far as the role of the Court and the environment is concerned, it is not surprising to say that appointments to the Court make a difference. More importantly, as Rosemary O'Leary has argued, while "[m]ost environmental conflicts never reach a court, and an estimated 50 to 90 percent of those that do are settled out of court," since the 1970s, the "courts in the United States have become permanent players in environmental policymaking" although their involvement in environmental affairs will "ebb and flow over the years."[9]

As we have seen in the discussion above, jurisdiction over environmental affairs has been divided between the major institutions of government. Once legislation is passed by Congress and signed into law by the president, executive agencies establish regulations as the lawmaking process places new responsibilities on the fifty states for implementing federal guidelines. However, as power has shifted from Washington to the states, subnational governments have not necessarily acted consistently in the implementation process. While some states have engaged in innovative efforts to improve environmental quality, others have opposed federal environmental guidelines or have not acted on federal legislation in a timely manner. Almost two decades ago, the research of Evan Ringquist clearly confirmed that the fifty states play an important role in environmental policymaking, James P. Lester reminded us that the actions of individual states are influenced by several factors including the state's wealth and the severity of its environmental problems compared to other states.[10] The value of Lester's work is that he organized states into policy types defined by their commitment to environmental protection and each state's institutional capabilities to take action. In doing so, a portrait of subnational government was established, placing the fifty states into one of four policy types.

As the Founding Fathers informed us in *The Federalist Papers*, the U.S. political system was created to control factions, yet the fragmented system of government also provides numerous access points for organized interests to pursue their causes, supported by the First Amendment right of free expression. Similar to other public policy issues, conflict over the management of the environment has resulted in a proliferation of ecological interest groups. Nonetheless, although these groups share a common commitment to protect the environment, they are characterized by different sociodemographic attributes, size, budget, tactics, and strategies. Moreover, not all ecology groups conduct themselves in politically legitimate ways. For instance, where the National Wildlife Federation is considered a "mainstream" organization that engages in influencing legislation or lobbying efforts, Greenpeace is identified as a "direct action" group whose members are willing to engage in nonviolent but confrontational actions (e.g., challenging whaling ships) while Earth First! has been characterized as a "radical" direct action group due to its willingness to go beyond nonviolent actions. In short, members

of environmental groups engage in conventional and unconventional participation, modes of behavior that will be further discussed later in this book.
Interest groups are not limited to the environmental movement. They have
ample options. The interests of business and industry are also represented
by a host of groups ranging from large groups with considerable resources,
including the American Petroleum Institute and the National Association
of Manufacturers, to smaller yet active groups that have focused on specific
or narrowly defined issues, such as the National Wetlands Coalition or the
Marine Preservation Association. Think tanks such as the Heartland Institute
and the Cato Institute push a conservative philosophy that is pro-develop-
ment/pro-growth, less inclined to support conservation efforts, and opposed
to federal and state environmental regulations on business and industry.

   While interest groups serve as linkage institutions that connect the
public to the political system, public opinion remains an important yet
problematic aspect of American politics. On the one hand, in a democratic
society the public's preferences should be expressed in government action.
Yet the extent to which this should be done is part of a long-standing
debate in American politics. How well is the citizenry informed about political and environmental issues? To what degree should policymakers rely on
public opinion as a guide for action? While some observers have argued
that the American public is not an informed, rationally thinking body of
individuals, others contend quite the opposite.[11] Although public opinion
data indicate that Americans hold strong views about environmental protection, to what extent do policymakers take these into account? Policymakers
must listen to their constituents but are also influenced by other political
and economic interests regarding their participation in the environmental
policymaking process.

   In addition to the role played by a variety of actors in American democratic politics, the United States also has an international role to play. The
United States is but one among some two hundred countries whose actions
affect the health of the planet, and it is a member of numerous regional
and international organizations that engage in environmental policymaking.
Similar to political conflict within the domestic policy arena, due, in part,
to differing interests, nation-states have shared, as well as distinct, concerns
that impact their orientation toward global environmental protection. For
example, the United States joined other countries and became a signatory
to the 1987 protocol addressing ozone depletion. In contrast, at the Earth
Summit in 1992, President George H. W. Bush didn't seek to unite the
United States with other members of the international community in their
effort to secure a global commitment to environmental quality. Although

global warming and biodiversity were salient issues at the summit, President Bush refused to sign the biodiversity treaty—the only participant to do so—and signed the global warming treaty only after it was revised to reflect voluntary rather than mandatory guidelines.[12]

Science, Politics, and the Environment

The environment is a policy area in which the well-being of the American people is determined by public officials at different levels of government. In the process of decision making, lawmakers are subject to numerous influences and, for example, they are inclined to reject scientific research that might be contrary to their self- or constituency interests, they might disregard what they don't understand, or they might hesitate to act when science lacks a consensus. Moreover, opposition to the scientific community can be found among politicians harboring ideological or partisan differences, business and industry leaders who are worried about their economic interests, citizens concerned about tax increases needed to resolve environmental problems, and state governments that might oppose the intervention of "big" government in the environmental policymaking process or are resentful at being forced to act due to unfunded mandates.

Against this background we are challenged by the following question: To what extent should science be involved in the environmental policymaking process? It has become commonplace to hear members of the scientific community argue that the earth's atmosphere, oceans, rivers, land, and wildlife have been profoundly affected by human activities. Some potential problems, among others, include increasing amounts of carbon dioxide released into the atmosphere, growing threats to global biodiversity through the destruction of natural habitats, reduced levels of clean water, and depleted supplies of the ocean's fisheries at a time when the human population is increasing. In short, how and in what ways do lawmakers respond to scientists who alert them to real and potential environmental problems?

As an example, global warming and climate change are key challenges for the scientist and lawmaker alike in the United States. On the one hand, 97 percent of climate scientists argue that the "greenhouse effect" is due, primarily, to human actions. On the other hand, entrenched economic interests, public officials guided by ideological rigidity, and a confused body of American citizens ensure that climate change will remain a divisive issue where inaction rather than progress carries the day. Moreover, the technical dilemma regarding decision making has a profound impact as one moves from the national to the global arena. As Lamont Hempel has argued, "Because attempts to solve global environmental problems invariably collide with the narrow self-interests of a state-centric system, few nations are prepared to follow the logic of collective environmental action to its political conclusion."[13] This does not deter action on the part of international political actors but it does make it more difficult.

Accordingly, in the United States, more needs to be done to ensure that the American public has a better understanding of science, and scientists must improve their understanding of and communication with American citizens.[14] This dilemma has been cogently described by Walter Rosenbaum in the following way:

> Public officials seek from scientists information accurate enough to indicate precisely where to establish environmental standards and credible enough to defend in the inevitable conflicts that follow. Scientists want government to act quickly and forcefully on ecological issues they believe to be critical. . . . The almost inevitable need to resolve scientific questions through the political process and the problems that arise in making scientific and political judgments compatible are two of the most troublesome characteristics of environmental politics.[15]

In the end, we face a fundamental challenge whether policymakers and the scientific community supported by the American public can establish effective measures to ensure appropriate responses to the myriad ecological problems that are salient at home and abroad.

Environmental Beliefs and Value Orientation

When discussing the politics of the environment, we are confronted with disputes over how to address environmental problems that are framed within value conflicts that occur between various stakeholders in the United States. During the late nineteenth and early twentieth centuries, the philosophical debate that occurred between John Muir and Gifford Pinchot—preservation versus conservation—set the stage for the future. For example, for Muir it was imperative to set aside public land in its pristine state for future generations. He articulated his vision of environmentalism in the following way: "Everybody needs beauty as well as bread, places to play in and pray in, where Nature may heal and cheer and give strength to body and soul alike."[16] Pinchot argued instead that land and natural resources could be used wisely and conserved for the future.

By the 1970s, a number of analysts set forth new explanations regarding the nature of values and value change in advanced industrial (postindustrial) democracies, including the United States. Samuel Hays, for instance, argued that as a result of post–World War II improvements in educational attainment and wider distribution of wealth in American society, new values took hold. According to Hays, "The driving force in the new interest in shaping improved levels of environmental quality were human and social values which took on an increasing level of importance in the second half of the twentieth century."[17] More importantly, Hays argued that "[t]he expression of environmental values and the evolution of environmental culture can be understood only in terms of its engagement with opposing values associated with development rather than environmental objectives."[18] One can easily imagine the preservation/development debate staged between advocates of yet another hotel in a row of hotels along a tourist beach and preservationists demanding that green space be maintained for today and tomorrow.

For more than four decades, Ronald Inglehart has conducted research about value change in postindustrial democracies.[19] Inglehart built upon the work of Abraham Maslow's hierarchy of needs, and he determined that, as a result of postwar prosperity and world peace, citizens' values were changing. Based on cross-national survey data, Inglehart reported that a new "postmaterialist" value orientation had emerged in which individuals put more emphasis on nonmaterial goals (e.g., a clean environment) than on materialist values (e.g., fighting rising prices).

Values and value change have an impact on public attitudes and behavior. In other words, values serve as "standards that guide conduct in a variety of ways."[20] Consequently, values (preservation, conservation, development) and value conflict affect our social and political outlook and influence the priorities of political institutions and the environmental policymaking process.

A Brief History of Environmental Policy in the United States

The Early Twentieth Century to the 1950s

During the first half of the twentieth century, the United States experienced periods of growth and depression, both of which militated against substantive governmental efforts to address the quality of the environment.

World War I, the Great Depression during the 1930s, World War II, and the Korean War turned the attention of political leaders to the issues of economics and national security. The period of the Roaring Twenties as well as the postwar prosperity during the 1950s created a mindset of unchecked growth, development, and continued exploitation of natural resources to meet consumer demands, industrial development, and national defense. Consequently, although environmentalists argued for years in favor of public policy initiatives to address issues ranging from the proper management of public lands and natural resources to resource depletion to air and water pollution, public officials tended to move incrementally rather than implementing a comprehensive national environmental policy.

Having said this, it is important to note that during the late nineteenth and early twentieth centuries, appropriate measures to manage the environment were promoted by several prominent individuals, including John Muir and Gifford Pinchot, and environmental groups (e.g., the National Wildlife Federation, Sierra Club)—measures that were the outcome of the continuing debate between conservationists and preservationists. For example, Glen Sussman and Mark Kelso have argued that the environmental measures that were advocated by the modern presidents beginning with Franklin D. Roosevelt were grounded in the conservation philosophy begun during the administration of Theodore Roosevelt:

> As Theodore Roosevelt moved the nation forward through industrial development and the politics of the Progressive era, he also had the vision to protect a large part of the nation's natural heritage by reserving huge tracts of public land for national parks, national forests, and wild preserves, embodying a conservationist strategy set forth by Gifford Pinchot, who would lead what we now know as the U.S. Forest Service.[21]

The conservationist philosophy of Theodore Roosevelt and Gifford Pinchot had a profound impact on American national politics. Not only did Roosevelt establish a model for his successors but the conservationist philosophy was "broadly accepted by Congress as well as the public and to a large extent extraction industries that were ensured access to resources."[22] Moreover, despite the fact that John Muir, president of the Sierra Club, promoted preservationist principles, Gifford Pinchot was successful in promoting the idea of conservation over preservation. As Pinchot argued:

> The first great fact about conservation is that it stands for development. . . . Conservation does mean provision for the

future, but it means also and first of all the recognition of the right of the present generation to the fullest necessary use of all the resources with which this country is so abundantly blessed. Conservation demands the welfare of this generation first, and afterward the welfare of the generations that follow.[23]

Consequently, Gifford Pinchot and President Theodore Roosevelt embraced the notion of conservation over John Muir's idea about preservation. In short, prior to the 1930s, the role of the federal government in environmental policymaking tended to focus on land management and conservation of natural resources.

The decade of the 1930s was characterized by both the expansion of the federal government generally and the increasing role of the federal government in environmental policy in particular.[24] During the era of Franklin D. Roosevelt, new and influential environmental groups were established, including the Wilderness Society (1935) and the National Wildlife Federation (1936). These groups began to exert pressure on political leaders, adding to the efforts already underway by groups such as the Sierra Club and the Audubon Society.

Moreover, the federal government became increasingly involved in environmental issues, in such initiatives as Franklin D. Roosevelt's Civilian Conservation Corps (CCC), the Tennessee Valley Authority (TVA), and the Soil Conservation Service. For example, the CCC played a significant role socially and economically by putting to work millions of unemployed young men, and environmentally through the planting of millions of new trees, fighting soil erosion, and protecting wildlife refuges. As a result of the Tennessee Valley Authority project, which provided much needed low-cost energy for American citizens, the environmental damage to the Tennessee Valley wrought by lack of planning was resolved and millions of trees were planted.[25] Also, the TVA was cited as attracting the attention of more foreign government leaders than any other resource conservation project, due to its success.[26] A. L. Owen described the conservation efforts of the 1930s in terms of the quality of planning: "The leadership necessary for the integration of any comprehensive plan was supplied by Franklin D. Roosevelt. Throughout his presidential years, he insisted upon the need for thoughtfully devised plans that would carry out an overall conservation policy."[27] FDR himself stated to the Congress as he began his first term in office in 1933 that programs like the CCC were

an established part of our national policy. It will conserve our precious natural resources. It will pay dividends to the present and future generations. It will make improvements in national and state domains which have been largely forgotten in the past few years of industrial development.[28]

At this time, Congress was instrumental in passing important environmental legislation that was signed into law by the president. These included the Taylor Grazing Act (1934), which addressed the problem of overgrazing on America's grasslands, and the Flood Control Act (1936), in which the U.S. Army Corps of Engineers assumed responsibility for implementing a policy to protect watersheds and improve flood control. The Roosevelt era also saw the United States engaged in several important regional environmental treaties that protected flora and fauna, including a treaty with Canada to protect salmon and halibut fisheries and a treaty with Mexico to protect migratory birds and animals.[29] During the early postwar period of the late 1940s and 1950s, presidents Harry S. Truman and Dwight D. Eisenhower were most concerned about national security issues and the communist threat rather than the environment. Although they issued several executive orders that were confined to land use and/or national parks and national forests initiatives, during a fifteen-year period, Congress passed and Truman and Eisenhower signed only a few pieces of significant environmental legislation. Moreover, Eisenhower had asserted that pollution issues should be considered a state and local responsibility rather than falling within federal jurisdiction.[30] James Sundquist has argued that the Eisenhower years were a time when "the federal government undertook few major new departures to conserve or improve the outdoor environment.[31]

Environmentalism: 1960s to the Present

The decades of the 1960s and 1970s were characterized by increasing levels of environmental initiatives by governmental authorities in the United States that involved presidential actions, congressional legislation, court decisions, state-level programs, and interest group activism, among others. In the early 1960s, for instance, biologist Rachel Carson moved the discussion about how to address new ecological issues of growing importance from the conservation-preservation debate to the environmental consequences of new technology. In her book *Silent Spring* (1962) she described the threat to the public and environmental health posed by increasing use of pesticides, especially DDT. She argued that "future generations are unlikely to condone our lack of prudent concern for the integrity of the natural world that supports all life."[32]

The Clean Air Act (1963) and the Clean Water Act (1972), for instance, were passed by Congress and signed into law by presidents Kennedy and Nixon, respectively. Subsequent clean air amendments were added in 1970 and again in 1977. Amendments were added to the Clean Water Act in 1977. The Endangered Species Act, which was passed in 1966, was amended and expanded in scope in 1969 and again in 1973. Although many other pieces of environmental legislation were passed by Congress and signed into law by the president, what was most significant was the increasing role the federal government began to assume in environmental policymaking. This was highlighted when both the government and the public embraced the first Earth Day in April 1970. That same year, the Environmental Protection Agency was created—a major development despite the failed effort to create a cabinet-level Department of the Environment and Natural Resources.[33]

At the same time, the judicial branch of government became increasingly involved in questions raised about the role of the federal government in environmental policymaking. As a result of congressional and presidential action, the National Environmental Policy Act (NEPA) passed into law and was signed by President Nixon in 1970. This compelled both the federal courts and the Supreme Court to respond to issues related to the scope and nature of NEPA in general and environmental impact statements (EIS) in particular.[34]

Furthermore, President Richard M. Nixon's New Federalism began a shift in responsibility for the implementation of federal environmental programs. When state and local governments had been responsible for environmental policy, priorities tended to favor development over preservation. As a result of changes at the federal level, states were becoming increasingly obligated to carry out environmental policy according to federal guidelines that also encouraged governors and state legislators to establish new subnational environmental initiatives and state-level environmental bureaucracies.[35] By the end of the decade, the Superfund Act (1980), which addressed hazardous waste sites and established a National Priority List for the most hazardous sites, was passed, as was the Alaska Lands Act (1980), which set aside millions of acres of land in the forty-ninth state.

During the 1980s, environmental protection was less a priority for the United States when Ronald Reagan assumed the presidency. The Reagan administration has generally been characterized as anti-environment, as it rejected previous bipartisan support for environmentalism and embraced instead a decidedly pro-development philosophy. Despite setbacks for several environmental issues including renewal of the Clean Air Act, which had sat dormant since 1977, Congress passed several pieces of legislation important to environmental protection, either with the signature of the president or by overriding his veto. Included among this legislation were the Safe Drinking Water Act (1986), Superfund Amendments (1986), and Clean Water Act Amendments (1987).

During the last decade of the twentieth century, only a few important environmental proposals were passed into law, namely, the Clean Air Act Amendments (1990) and the California Desert Protection Act (1994). Notwithstanding former president George H. W. Bush declaring himself the "environmental" president and the fervent hope among environmentalists that Bill Clinton would be a "green" president, little substantive action occurred. Bush used the resources of the presidency to ensure passage of the Clean Air Act amendments. However, he disappointed environmentalists when he reversed his position on environmental issues, in response to pressure from fellow Republicans in the Congress and business and industry interests. The California Desert Protection legislation was successful due to the efforts of California's senators, primarily Dianne Feinstein. Still, those who supported and worked for the legislation were bolstered in their efforts, knowing that they had an ally in the Clinton White House who would at least sign rather than reject the bill.

In the Congressional elections of 1994, the Republicans captured both chambers of Congress for the first time in four decades. Faced with an obstructionist Republican-majority Congress, Clinton used the 1906 Antiquities Act in order to set aside large tracts of public land. He did so in 1996 when he established the Grand Staircase-Escalante National Monument in Utah despite local opposition to his action. As he reached the end of his presidency, Clinton set aside millions acres of land—an effort to bolster his environmental "legacy." Clinton was attempting to act as a "conservationist" president following in the steps of Theodore Roosevelt by preserving public lands for future generations.

Clinton's successor, George W. Bush, the first president of the twenty-first century, made it clear from the outset of his administration that he would follow and expand the pro-development, anti-regulatory approach set forth by Ronald Reagan. This was demonstrated early in his presidency when he dealt with three important issues—namely, water quality, oil exploration, and carbon emissions. Partly in response to President Bill Clinton pushing for stricter standards regarding the amount of arsenic in drinking water, Bush indicated that he would relax the standard. Bush and his EPA were eventually compelled to comply with the Clinton standard due to pressure from Congress and the public. Bush spoke frequently about opening Alaska's Arctic National Wildlife Refuge for oil and gas exploration. He argued that this would be a way to achieve energy independence. However, he was opposed by members of the Congress and environmentalists who were concerned about protecting wildlife and a pristine environment. As his first administrator of the Environmental Protection Agency, Christine Todd Whitman informed us that during the presidential campaign of 2000 the Republican presidential candidate was committed to reducing carbon dioxide emissions that are associated with the greenhouse effect, global warming, and climate change. Once in office, however, Bush remained an opponent of substantive action to address global warming and climate change. Later in his presidency, Bush used the power resources of his office in support of efforts opposed to environmental protection.[36] He offered his Clear Skies Initiative, which would weaken the Clean Air Act, and he pushed his Healthy Forests Initiative that would make millions of acres of forests exempt from environmental review. He attempted to weaken the Clean Water Act by allowing mining companies to be exempt from compensation when wastes polluted waterways, wetlands, and streams. In short, during his presidency, not one major piece of environmental legislation was passed.

As a result of the election of Barack Obama in 2008, Democrats regained control of the White House and the Senate. However, two years later, the Republicans regained control of the House of Representatives. This set the stage for partisan executive-legislative contention over environmental policymaking. While the new president would be challenged by a host of issues ranging from a national crisis facing financial institutions to a weakened auto industry to home foreclosures to two wars, he offered hope to environmentalists who were thrilled by his election. Obama engaged in a number of efforts to include the environment as part of his larger public agenda. During his first term in office, he signed the Omnibus Public Lands Management Act that would set aside more than two million acres of public land as national wilderness. He issued an executive order that would commit state and local governments to work with the federal government in an effort to maintain the health of the Chesapeake Bay, the largest estuary in the country. He also announced a National Fuel Efficiency Policy that would impose increased fuel efficiency standards on new vehicles and at the same time cut greenhouse gas emissions. Nonetheless, he has been criticized for having a weaker record than George H. W. Bush, Bill Clinton, and George W. Bush when it came to cleaning up toxic waste sites. Moreover, Obama's Department of the Interior and the Fish and Wildlife Service failed to terminate a war on America's grey wolves that continues in different parts of the country.

While this discussion so far has been focused primarily on domestic environmental affairs, in the international arena, the United States has also engaged in several important global initiatives. For example, with the support of President John F. Kennedy, the ratification of the Limited Nuclear Test Ban Treaty with the Soviet Union in 1963 moved the two adversaries away from potential nuclear conflict and toward mutual nuclear arms control. It also reduced the public health and environmental risk posed by radioactive debris resulting from above-ground nuclear testing. A decade later, the International Convention on Trade in Endangered Species of Wild Fauna and Flora (CITES) was a global effort to protect endangered plants and animals. The United States was the first nation to ratify this treaty in the mid-1970s, prohibiting international trade while promoting conservation of flora and fauna. The treaty was ratified by nearly one hundred nations by 1987.[37] Also in 1987, President Reagan signed and the U.S. Senate ratified the Montreal Protocol on Ozone Depletion. The accord was an important expression of the multilateral effort to address "new" global climate environmental issues. Reagan's successor, George H. W. Bush, used the resources of his office in support of the 1990 Clean Air Act amendments that improved relations with Canada over acid rain caused by emissions from power plants in the United States. At the same time, Bush supported the Earth Summit's commitment to addressing greenhouse gas emissions as long as the effort was voluntary not mandatory.

The North American Free Trade Act (NAFTA) signed into law in 1994 and geared toward enhancing free trade, was supported by President Bill Clinton. Although environmental groups voiced their concern about the ecological impact of the treaty, Clinton stressed the importance of environmental protection via provisions added to the agreement. Seven years later, only two months into his presidency and despite a campaign pledge to reduce carbon emissions, George W. Bush rejected the U.S. commitment to reduce greenhouse gases associated with global warming and climate change. In contrast, Barack Obama, the forty-fourth president, has had a promising record when considering the issue of climate change. He has been involved in several efforts, both regional and international, in support of clean energy technologies. Having said that, he has been strongly opposed by House Republicans in his effort to make progress on a climate change agenda.

For more than a half-century, the history of environmentalism in the United States has been characterized by conflict and compromise as the federal government increased it role in environmental management. During this period, the environmental policymaking process has involved a variety of old and newly emerging ecological issues. For example, environmental policymaking was, during the late nineteenth century, first concerned with the conservation of public lands. Since then, American citizens have been confronted with the changing nature of environmentalism and the evolution of environmental problems, namely, the first-generation problem of air and water quality; second-tier issues including toxic and hazardous wastes; then, new, third-generation issues involving stratospheric ozone depletion, global warming and climate change, and biodiversity.[38] Global warming and climate change, in particular, have constituted a quite different range of issues for American citizens since, in contrast to the visibility of air and water pollution, the nature of a global "greenhouse effect" is difficult to grasp and quite remote from the typical person's realm of understanding.

Design of the Book

The environment as an important public policy issue in the United States is the focus of this book. Its purpose is to assess the roles of both political institutions and the public in the making of environmental policy and to offer the reader insight into how the American political system works. The book includes several features unique in the study of U.S. environmentalism. First, we use an institutional/behavioral approach—namely, how do institutions and the political actors working within them respond to environmental problems? In doing so, in contrast to other books that focus on specific environmental issues in each chapter, we turn our attention to politics and the political process. Second, we include two box inserts in each chapter that focus on a person and a case study. The person and the case study are linked to the institution being covered in the chapter. Third, we include a box insert in each chapter that focuses on global climate change. This is an innovative mechanism that ties the chapters together. Finally, we provide a set of questions in the preliminary discussion of the chapters below that guide the analysis that begins with chapter 2.

Chapter 1 provides an analytic framework for the chapters that follow by discussing how the organization of American constitutional democracy influences the policymaking process. In doing so, it narrows its focus to environmental politics and policy, including how government and policymakers shape environmental policy. The chapter also provides a discussion of science and politics important to environmental policymaking and gives attention to the role of environmental belief systems and value orientation that impact the policy making process. In short, each chapter will focus on a single institution (two related institutions in chapter 3) and examine major environmental debates, decisions, accomplishments, and problems.

American federalism, intergovernmental relations, and the environment are examined in chapter 2. The discussion in the chapter analyzes the historical roots of relations between the national government and the fifty state governments generally, and the contemporary dynamics of federal/state relations in the shaping of environmental policy. This will be conducted against the backdrop of the devolution of power from Washington to the state capitals. How important have state actions been in shaping national policy? Why might some states be active in promotion of environmental policies while others have been resistant to the same? Which states, if any, have initiated creative environmental policies?

Chapter 3 evaluates the impact of public opinion and interest groups on environmental issues. Public opinion polls provide a portrait of American citizens' attitudes about a host of environmental issues. Yet government action does not always reflect public preferences. Interest groups serve as an important linkage institution that ties the American public to the governmental process. How important has the public response to the environment as a policy issue been in the shaping of environmental policy? How does one explain variation in public opinion about environmentalism? What has been the pattern of public opinion over time? Which interest groups have been most influential over the years in shaping environmental policy? What kinds of tactics and strategies have environmental groups employed in the promotion of environmentalism?

Congress, the legislative process, and the environment are addressed in chapter 4. As a deliberative body engaged in the process of bargaining and compromising among diverse interests, Congress can either work with the president or compete with the president's goals. Congressional efforts in environmental policymaking have been characterized by bipartisanship as well as partisan differences. What is the nature of the legislative process in creating environmental policy? Which committees and congressional leaders have been most influential in shaping environmental policy? How has partisanship united or divided members of Congress when voting on environmental policy? Which Congresses have been most productive in producing environmental policy and what are the key pieces of environmental legislation passed by the Congress?

The environmental presidency is the focus of chapter 5. Although the president is the most visible political figure in American politics, the chief executive is confronted with a diversity of public policy issues, among them environmentalism. The level of presidential action depends, of course, on a variety of factors. The roles played by the president (e.g., legislative leader, environmental diplomat) help explain presidential involvement in environmental policymaking. Has the environment been at the center of the president's public agenda? How has the presidency compared to the Congress in the promotion of environmental protection? Which presidents have been more protective of the environment and which presidents have promoted a pro-development philosophy toward the environment? Which presidents, if any, can be considered "environmental presidents"?

We examine the executive branch of government and the role played by executive agencies and environmental policy in chapter 6. The federal bureaucracy comprises numerous agencies, bureaus, departments, and commissions with jurisdiction over the environment and each has varied in terms of its influence on environmental policymaking. What has been the role of presidential influence and the independence of executive agencies in the shaping of environmental policy? What has been the impact of key personnel and/or heads within the executive bureaucracy in environmental policymaking? Has the Environmental Protection Agency or the Department of Interior been the dominant player in environmental policy making? Which other executive agencies have been important in environmental policymaking?

The environmental court is the focus of chapter 7. Similar to its two federal counterparts, the judiciary has played an important role in the life of the nation and environmental issues. How important have Supreme Court decisions been in shaping environmental policymaking? How important have the Court's decisions been in influencing other policymakers? How have other political actors in the polity responded to the Court's decisions? How influential have individual justices been in particular cases involving environmental decision making?

Chapter 8 concentrates on global environmental politics and policy. While most of the discussion in this book addresses domestic politics, the United States also has a role in the international environment. Regional and international treaties have been signed and are in force, and regional and international organizations have increasingly included the environment as a policy area demanding global attention and solutions. What have been the major global environmental issues? What are the major international organizations and nongovernmental organizations (NGOs) involved in global environmental policy? How successful have international environmental agreements (treaties, protocols) been in protecting the environment? What is the relationship between national security and global environmental security?

The concluding chapter evaluates the U.S. approach to environmentalism. In doing so, it assesses how political institutions and policymakers have responded to environmentalism at home and abroad. In the discussion leading up to this concluding chapter, we examine how American political institutions and the individuals working within each of them have shaped environmental policymaking. Based on our observations, we close with a set of propositions that offer the reader a better understanding of American politics and the environment.

The discussion that follows examines the environment from the perspective of the various policy units in the political system. As you, the reader, examine the role played by each institution covered in the following chapters, keep in mind how political actors responded to environmentalism within the institutional setting. How might the political behavior of citizens and public officials be characterized in analyzing environmental policymaking? What have been the major influences on political institutions and the political actors working within them? Why have some policy actors embraced the effort to protect the environment while others have resisted or delayed environmental initiatives? Finally, consider to what extent the environment, in comparison to other public policy issues, has been an important issue in American politics.

2

American Federalism and
Environmental Politics

The U.S. Constitution separates power among executive, legislative, and judicial branches of government and also between the national government and states. The term *federalism* is generally used to describe constitutional division of powers between the national and the state governments. In a federal system, policy guidance may come from the president and Congress, but policy implementation must be responsive to input from the states and localities, as well as from national officials. Naturally, conflict is not unusual when different power centers are involved in policymaking. This is especially true in environmental politics, where there have been widespread calls for devolution of regulatory authority to subnational governments in recent years.

In this chapter we examine environmental politics in the context of a federal system and intergovernmental relations. Historical and contemporary working relationships and linkages among key institutional players will be examined. Implementation strategies using regulatory "sticks" and nonregulatory "carrots" will also receive attention. Finally, a personal profile and two issue-based case studies will highlight various dimensions and dynamics of federalism, leadership challenges, and environmental politics by focusing on controversial disputes and collaborative actions arising in particular states.

Federal-State Relations: Historical Roots

The distribution of power between the national government and states and localities has shifted back and forth over the years. When states failed to adequately address problems that spilled across their boundaries, the national government often intervened. Recently, a reverse trend occurred, with a shift in power back to the states and localities for managing certain environmental programs. Six historical periods in the evolution of national environmental policy are briefly summarized below.[1]

The Common Law and Conservation Era: Pre-1945

Early in our nation's history, selling federal lands was a source of revenue, and federal land grants spurred states to pursue natural resource development projects. Federal-state conflicts during this period were infrequent because federal development policy coincided with state preferences. The late 1800s witnessed a shift of emphasis from development to conservation. This change led to growing discord because states and private parties, who were accustomed to federal policies promoting development and facilitating transfer of public lands, were now subject to new constraints linked to conservation. Conservationists believed in the sustainability of natural resources; preservationists sought to preserve wilderness areas from all but the most limited uses (e.g., education, recreation).[2]

Environmental regulation was limited in the pre–World War II era primarily to local policies protecting public health, and infrequently to state policies. The inability of the states to respond to the demands of the Great Depression, and increased federal regulation with the advent of the New Deal, altered the conception of federalism. That is: national problems require national responses. Hence, in the post-1945 era, the federal role in environmental protection policy increased.

Federal Assistance for State Environmental Problems: 1945–1962

For almost twenty years after World War II, the federal government promoted environmental protection by providing research and financial assistance to states and localities. As industrial pollution crossed state lines, the federal government needed to act, but actions were designed to encourage and assist the state's ability to respond. In the 1950s and 1960s, congressional funding for state water and air pollution control, and municipal sewage treatment plants was predicated on the idea that state and local governments bore responsibility for addressing environmental problems. Nonetheless, during this period, federal policymakers were increasingly aware of the national scope of such pollution problems. At the same time, new environmental organizations emerged and existing ones expanded.

The Rise of the Modern Environmental Movement: 1962–1970

Publication of Rachel Carson's *Silent Spring* (1962), warning the public about the dangers of pesticides, marked the beginning of the third era; a warning about the population explosion followed with publication of Paul Ehrlich's *The Population Bomb* (1968). Congressional policy debates pitted those supporting development of public resources against those favoring environmental protection, especially focused on balancing recreation and economic interests against environmental concerns in areas such as national forests. Legislative initiatives primarily targeted federal agency actions, rather than private sector activities, seeking regulations to ensure that government projects took into account environmental concerns. Three other important developments in this period include increased judicial attention to government agency actions regarding the environment, emergence of new environmental interest groups, and passage of legislative acts such as the Clean Air Act, Water Quality Act, Endangered Species Conservation Act, and the National Environmental Policy Act (NEPA) of 1969.

Erecting the Federal Regulatory Infrastructure: 1970–1980

The decade of the 1970s witnessed a legislative explosion, with twenty important national environmental laws passed or significantly amended. This resulted in an expanded federal role in environmental protection, establishment of new federal standards and program requirements, continued attention to parks and wilderness, and enhanced access for citizen activists to voice their concerns via administrative procedures and litigation. Also, by executive order in 1970, President Nixon created the U.S. Environmental Protection Agency (EPA). While several laws adopted in this period imposed national standards and provided for substantial national regulation, most delegated significant authority to the states for program implementation and enforcement.

Extending and Refining Federal Regulatory Strategies: 1980–1990

The 1980s saw passage of additional legislation as well as augmentation of statutes passed earlier. Some laws relied heavily on states for implementation, but delegation by this was sometimes slow and uneven. For example, the Comprehensive Environmental Response Compensation, and Liability Act ("Superfund") created a program for cleanup of hazardous substance releases, authorizing states to make decisions on cleanup; however, it took more than ten years for the EPA to follow up and actually delegate cleanup decisions to the states. Nonetheless, rigid implementation guidelines with consequences for inaction were found also in amendments to several environmental laws. Other developments during this decade include mobilization by industry to curb the growth of environmental legislation, greater reliance on administrative decision making rather than legislative action, lax enforcement of federal environmental standards, and reduced federal funding to implement mandates.

Regulatory Recoil and Limits on Federal Power: 1991-Present

The early- to mid-1990s witnessed a decided change in the views of national environmental protection policymakers, with Congress and the president seeking to reduce regulatory burdens on states and localities, and the Supreme Court questioning long-held views on national regulatory authority. For example, President Clinton's Executive Order 12875 in 1993 and congressional passage of the Unfunded Mandates Reform Act of 1995 both sought to make it harder to impose unfunded mandates on states and local governments. Court decisions addressed the question of constitutional limits on federal regulatory authority and indicated that the Court was reconsidering its long-standing views on federalism. Furthermore, difficulties reauthorizing existing federal environmental laws (e.g., the Endangered Species Act), or efforts to weaken them, signaled a new and less favorable sentiment in Congress regarding environmental matters.

Contemporary Federal-State Interactions:
Implementation and Enforcement

Legal relationships between federal and state governments take various forms.[3] *Delegated* programs authorize a federal agency such as the EPA to establish national standards and charge the states with primary implementation and enforcement responsibilities. *Voluntary* programs provide inducements ("carrots") to states without authorizing federal agencies to manage programs inside state jurisdictions. *Mandated* programs impose requirements ("sticks"), by federal law, on states. In the process of implementation and enforcement, governments use a wide variety of regulatory "sticks" and non-coercive regulatory "carrots." Among the sticks are command-and-control regulation, oversight, technology-based requirements, permits and inspections, enforcement, and unfunded mandates. Each of these involves use of coercive powers of higher levels of government to influence the behavior of those at lower levels.

Sticks

COMMAND-AND-CONTROL

Conventional regulation relies on national environmental standards with substantive and procedural requirements, tight timetables, inspections, controls, penalties for noncompliance, and litigation. Grants-in-aid programs are funded by the national government and given to state and local governments on the condition that the monies be used for purposes specified by the federal government. Crossover sanctions refer to actions by the federal government to withhold funding in a broad range of programs when lower governments fail to perform in any specific program. Critics of command-and-control regulation contend that it is costly, narrowly focused (e.g., on one pollutant or point source polluting activity), inefficient, inflexible, fragmented, concerned with remedial action rather than pollution prevention, adversarial, cumbersome, and slow to respond to changing conditions.[4]

OVERSIGHT

After laws are passed and regulations issued, legislative and bureaucratic oversight begins. Higher-level government routinely oversees the compliance activity of lower-level governments, just as government at all levels oversees compliance activity of businesses and nonprofit organizations. Subnational administrators often follow orders and instructions from federal officials, although those sophisticated in bureaucratic politics have in some instances been able to successfully resist or modify policies and guidance from higher-level officials. Some effective methods of oversight include federal investigation, audits, and reviews of state action; reversion of certain decisions to federal officials in cases involving major impact; suspension of state action if federal authorities object to application of program guidelines; and revocation of state authority due to noncompliance with federal conditions.[5]

TECHNOLOGY BASED

Environmental laws often contain phrases regarding the required technology to be used in pursuit of national policy goals. For example, the 1990 Clean

Air Act established very specific and precise standards for technology, while other laws include more ambiguous phrases such as "best available technology." Still others may have less stringent standards. Strict, precise mandates intentionally restrict the discretion and flexibility of those implementing the law and regulations. All of these mandates have substantial cost implications for those subject to their statutory requirements. The "stick" helps to ensure that the intent of national policy is followed, but it imposes burdens on those charged with responsibility for carrying out or complying with the mandate.

PERMITS AND INSPECTIONS

Alternative permitting systems and revamped inspection practices have been tried in various states. Certain states have strengthened penalties and sanctions, while others have loosened controls and reengineered processes. For example, Missouri streamlined the application process for stormwater construction permits through its ePermitting system, which saves time for applicants and staff. South Carolina developed an interactive Permit Central website enabling customers to determine which permit they will need and how long it will take.[6] Accountability is maintained via annual reports certifying compliance, increased inspections and audits, vigorous enforcement against noncompliant firms, and other methods. Other state reforms have sought to help permit applicants by creating advocacy or permit-assistance offices, streamlining procedures, and encouraging prompt decisions.

ENFORCEMENT

The federal government has traditionally relied heavily on enforcement mechanisms which are thought to deter polluters. State governments have taken steps to bolster enforcement of environmental laws as well. For example, New York has created a multifaceted plan that includes, among other things, a Comprehensive Enforcement Team, regional enforcement coordinators, heavy reliance on inspections, and a new air enforcement unit.[7] The EPA's oversight and enforcement role is complicated by the variations in state organizational structure. For example, state environmental protection programs may be housed in public health agencies, mini-EPAs, environmental superagencies, or in separate boards or commissions.[8]

UNFUNDED MANDATES

Unfunded federal mandates involve obligations imposed on states by the federal government without monetary compensation. As mentioned previously, congressional response to state and local government concerns about those escalating costs was passage of the Unfunded Mandates Reform Act of 1995. Under this law, unfunded mandates in excess of $50 million annually must be identified and separately voted upon. Costly regulatory requirements ($100 million or more annually) must be identified as well. Only five rules over the past ten years required public sector mandates on subnational governments; the EPA issued three of these.[9] Unfunded environmental mandates pose numerous problems for subnational governments, including "fragmentation (institutional, scientific, legal and political), lack of information, and rigidity of laws and regulations."[10] The outcry from financially strapped state and local governments has altered the nation's approach to environmental protection.

Carrots

"Carrots" or noncoercive strategies used to influence the behavior of subnational governments and firms include voluntary compliance, public education, preventive efforts, technical assistance, market development, market-based approaches, privatization, partnerships, and user charges and tax policies. In many instances, federal funding provides the stimulus (carrot) for state programs to pursue such strategies. Together with the more coercive strategies discussed above, these tools are part of the strategic arsenal of higher levels of government and will be discussed in turn.

VOLUNTARY COMPLIANCE

In some instances, regulatory "sticks" are not necessary; once organizations know what is required of them, they may willingly comply. In order for voluntary compliance to be effective, firms and other organizations must be aware of what is expected and have the incentives and resources to comply. In California, the Air Resources Board provided pollution control district inspectors with technical manuals, and industries were given free simplified handbooks about air quality standards (see profile of Mary Nichols and CARB below). In Maine, the Sustainability Commission provides technical assistance for firms and people in waste reduction, recycling, and composting; similar assistance is provided in Minnesota to aid in cleanup of contaminated urban land.[11]

Personal Profile

Mary Nichols and CARB: Environmental Champions

The United States currently lacks comprehensive legislation regulating greenhouse gas (GHG) emissions. The inability of the federal government to do more on climate change has helped spur states to take action.[1] California has been a leader among the states in innovative environmental policy. The California Air Resources Board (CARB), under the leadership of its chairperson Mary Nichols, is paving the way by implementing AB 32, California's Global Warming Solutions Act of 2006, pathbreaking comprehensive state climate legislation. Ms. Nichols oversees the design and implementation of programs under this first-of-its kind landmark legislation.

Mary Nichols draws on rich experience to inform decisions in her pioneering leadership role. In her youth she was an activist on peace and civil rights issues. Later, she was educated at Cornell University (BA) and Yale Law School (JD). Early in her career she worked as a journalist at the *Wall Street Journal*, an attorney with the Center for Law in the Public Interest bringing cases on behalf of environmental and public health organizations, assistant administrator at the USEPA working on air and radiation program, and California's Secretary for the Natural Resources Agency. Her public service posts also included executive director of Environment Now Foundation, founder of the Office of Natural Resources Defense Council in Los Angeles, and UCLA professor and director of the Institute of the Environment. She has served three terms as chair of CARB, having been appointed by Republican Governor Arnold Schwarzenegger and twice by Democrat Jerry Brown, a testament to her evenhanded, bipartisan approach.[2]

Her contributions to the environment have been widely recognized in high-visibility profiles: recognition as one of the World's 100 Most Influential People by *Time* magazine, one of 12 Leaders Who Get Things Done by *Rolling Stone*, Personality of the Year by *Environmental Finance* magazine, and Queen of the Green in *Dan Rather Reports*.[3] Indeed, she has been called the "Thomas Edison of environmentalism" for her fierce commitment to innovative technology and commonsense approaches that provide a model for her state, nation, and the world.[4] As EPA Administrator Lisa Jackson says, "She's a game-changer, because she knows how to put policies into action so they stick."[5] Among her priorities are implementing the AB 32 climate change program, working with CARB to reduce diesel pollution at ports, and approving performance-based regulations directed at cleaner air. Her successes in California and Washington, D.C., have helped to boost fuel economy, cut acid rain and greenhouse gases—setting standards applicable globally.[6] An advocate for environmental justice, Nichols and CARB take pains to examine decisions with care to ensure that strategies are sound and safeguards are in place to avoid worst-case scenarios that disproportionately harm low-income communities.[7]

The world's second-largest comprehensive cap-and-trade program was adopted by CARB in 2011 to harness market forces and curb global warming. It was designed and put into effect by Nichols and CARB with 2013 as the first compliance year. Implementation efforts quickly faced a two-front attack—a legal challenge and a statewide referendum seeking to kill the cap-and-trade program legislatively. Nichols was able to fend off the attack and preserve the climate plan.[8] The program sets lower limits on companies' GHG emissions and allows those who emit less than their cap to sell permits to those who exceed their limits.[9] This was one prong of the strategy for implementing AB 32. Cap-and-trade together with innovative rules and regulations and investment in new technologies has helped to curb California's GHG emissions. The AB 32 goal, once thought unattainable, of reaching 1990 emissions levels by 2020 now seems more realistic. A longer-range goal is 80 percent below 1990 levels by 2050. Twenty percent of the 2020 emission reduction goal is to be attained with cap-and-trade.[10] Putting a price on carbon is fundamental to the strategy. In cooperation with the USEPA Mary and CARB are also seeking to achieve a zero-emissions automobile fleet.

Mary Nichols is a politically savvy, technically proficient, and cost-conscious public servant who is making a difference as chairperson of a state-level board authorized to exercise government powers to achieve ambitious environmental and public health goals.

Notes

1. Andrew H. Meyer, "A. Navigating Climate Regulation on Dual Tracks: Experts Discuss the Challenges and Pitfalls of AB 32 and the Clean Air Act," *Clean Energy Law Report*, March 18, 2014; http://www.cleanenergylawreport.com/environmental-and-approvals/navigating-climate-regulation-on-dual-tracks-experts-discuss-the-promises-and-pitfalls-of-ab-32-and/. Accessed September 23, 2014.

2. Juliet Eilperin, "California Establishes Carbon Market," *The Washington Post*, December 17, 2010; http://voices.washingtonpost.com/post-carbon/2010/12/california_establishes_carbon.html. Accessed September 23, 2014.

3. "Mary Nichols, Chairman, California Air Resources Board," *California Environmental Protection Agency: Air Resources Board*, last modified October 10, 2013; http://www.arb.ca.gov/board/bio/marynichols.htm. Accessed September 23, 2014.

4. Lisa P. Jackson, "The 2013 TIME 100: Pioneers, Mary Nichols," *TIME*, April 2013; http://time100.time.com/2013/04/18/time-100/slide/mary-nichols/. Accessed September 23, 2014.

5. Tim Dickinson and Julian Brookes, "The Quiet Ones: 12 Leaders Who Get Things Done," *Rolling Stone*, January 2012, 31–34.

6. Jackson, "The 2013 TIME 100."

7. Mary Nichols, interviewed by Agustín F. Carbó-Lugo, Yale Law School, New Haven, CT, February 25, 2012.

8. Dickinson and Brooks, "The Quiet Ones."

9. Dana Hull, "Mercury News Interview: Mary Nichols of the California Air Resources Board," *San Jose Mercury News*; http://www.mercurynews.com/ci_21960742/mercury-news-interview-mary-nichols-california-air-resources. Accessed September 23, 2014.

10. Tauna M. Szymanski, "California Carbon Market: Challenges and Opportunities," *UCLA School of Law*, March 2014; http://ali-cle.org/index.cfm?fuseaction=online.chapter_detail&paperid=296968&source=2. Accessed September 23, 2014. For more information on CARB see Ann E. Carlson, "Regulatory Capacity and State Environmental Leadership: California's Climate Policy," *Fordham Environmental Law Review* 24, no. 1/2 (Spring 2013): 63–86 and Ann E. Carlson, "Iterative Federalism and Climate Change," *Northwestern University Law Review* 103, no. 3 (2009): 1097–1161.

PUBLIC EDUCATION

Public education, outreach, eco-information programs, energy efficiency labeling, and participation comprise strategies designed to improve environmental awareness and pollution prevention. States have developed innovative policies to bolster participation in high-stakes public hearings using online informational meetings and Web-based comment forms (e.g., Oregon) and to protect public health using a website including data and mapping tools to inform about remediation sites (Indiana).[12] Another example of the benefits of such activities is the information disclosure of toxic releases, which provides a measuring rod for stakeholders to assess the pollution records of various manufacturing firms.

PREVENTIVE EFFORTS

The federal government and several states have tried to refocus attention from regulating pollutants after they appear to preventing pollution before it occurs, but changing to this method has been slow. For example, Massachusetts' Environmental Results Program seeks to prevent pollution by certifying compliance with requirements, creating performance-based standards, providing technical assistance, and conducting compliance audits.[13] These preventive efforts are raising awareness and improving environmental protection. State prevention initiatives in Massachusetts and New Jersey have benefited from federal grants and require facilities to project reduction targets for each covered toxic pollutant.[14]

TECHNICAL ASSISTANCE

Federal policymakers have broadened their concern in the past two decades from focusing primarily on environmental cleanup to sponsoring efforts to minimize waste, conserve energy, and prevent pollution. This shift is evident in the passage of the Pollution Prevention Act of 1990. This law authorizes the EPA to give small grants to assist states in offering technical assistance in pollution prevention. The previously mentioned Minnesota and Massachusetts laws go far beyond conventional technical assistance in their comprehensive toxic pollution prevention programs.

MARKET DEVELOPMENT

Some states are creating new markets that have positive environmental effects. In California, environmental protection was improved when an agreement between the state and Atlantic Richfield (ARCO) led ARCO to acquire land mitigating the impact of oil wells on endangered species in exchange for permits to drill additional wells. ARCO then can sell land bank acreage to other developers with environmental obligations. In Florida, state and local water management districts run wetlands mitigation banks.[15] In Pennsylvania, a gubernatorial task force has created a market for recyclable materials by establishing an electronic bulletin board and hosting conferences of buyers and sellers of recyclable products. Each of these innovative undertakings develops a new market.

MARKET-BASED APPROACHES

Market-based tools and strategies are advocated by some as an alternative to traditional regulation. The conventional command-and-control approach sets uniform mandates, typically in the form of technology or performance standards. The market-based approach seeks to influence behavior using pricing mechanisms instead of specific standards for levels or methods of pollution control. This enables environmental goals to be achieved without complex legislation, especially in the areas of waste management, land use, and air quality. Examples include unit pricing for solid waste collection and disposal that establishes a direct connection between the amount of waste generated and prices charged, and land trading systems, or tradable permit programs, whereby governments license owners who preserve or upgrade their property and then authorize them to sell credits to developers.

PRIVATIZATION

Government at all levels is increasingly contracting for services with the private and nonprofit sectors. A potential danger in privatizing government services is the loss or reduction of public accountability. A distinction can be made between "formal" and "informal" privatization with national policy objectives more likely to be met under more formalized privatization arrangements.[16] Examples of privatization in the environmental protection area are somewhat limited. However, privatization initiatives are frequently found in the areas of solid and hazardous waste collection and disposal.

PARTNERSHIPS

Partnerships are a means to link knowledge, experience, and resources to address environmental problems. The National Environmental Performance Partnership System (NEPPS) was an EPA initiative to facilitate identification of national, state, and local priorities and target resources to address them. The mechanisms to achieve this include partnership agreements and grants designed to enhance flexibility and reduce federal oversight. The EPA has established core performance measures to aid in implementing priorities and strategies and to guide development of work plans and agreements. This has increased flexibility and reduced the reporting burden put on the states. In addition to federal-state partnerships, states are negotiating public-private partnerships, multiagency partnerships, multistate working groups, and agreements with professional associations, Native American nations, and local governments.

USER CHARGES AND TAX POLICIES

Charges to citizens and organizations that actually use a service are not uncommon in the energy and environmental policy area. One advantage of user fees and taxes is the heightened awareness that citizens gain regarding the cost of such matters as water usage and waste disposal. Minnesota relies on user fees from participating organizations to cover the administrative expenses of its Voluntary Investigation and Cleanup program of contaminated lands. More than four hundred state environmental taxes, fees, and surcharges have been adopted, including hazardous waste disposal fees, underground storage tank fees, and waste tire fees, among others. The resulting revenues in many cases are used to cover the cost of programs such as recycling and renewable energy.[17] It is a common practice for local governments throughout the country to use unit pricing (by the bag or can) for trash pickup, sometimes called "pay-as-you-throw" garbage programs. Also, several states provide tax refunds on beverage containers or tax credits for purchase of recycling or renewable energy equipment.

Attention now shifts from the historical evolution of federalism and the environment, as well as the sticks and carrots used in federal regulation of environmental matters, to two case studies, which show the political dynamics involved in policy implementation and enforcement in mountaintop removal in West Virginia and sea-level rise in Southeast Florida.

Case Study

The Dynamics of Implementation and Enforcement I: Mountaintop Removal in West Virginia

Environmentalists and coal mining interests in West Virginia have for many years been embroiled in a heated dispute related to a process called "mountaintop removal" strip mining. West Virginia is a major supplier of the country's valuable, low-sulfur coal; as a nation we rely on coal for most of our electricity. Mountaintop removal is the method used to extract this coal. It involves use of explosives and huge equipment to blast away entire mountaintops, uncover the coal—which is then scooped up and hauled away—and, finally, dump the leftover rubble and dirt into the valleys, hollows, and streams below. Congress outlawed the practice in 1977, but it continued until action by a federal judge found that the West Virginia coal mining industry was breaking the law.

In the process of mining coal, the environment is adversely affected in numerous ways: by decapitating mountains, burying streams, destroying grasslands, and stripping off topsoil. Area residents complained that flying debris damaged their homes. Aggrieved parties filed suit against the mining industry, contending that operators violated state and federal laws and that regulators have been lax in enforcing these laws. Court decisions have banned coal mining companies from dumping waste into streams, but over time such decisions have been modified. This case of mountaintop removal illustrates the intimate connection between state and national politics, and the ways various interests can seek to have their voices heard through different institutions in the political process. It also shows the potential conflict in many jurisdictions between economic concerns and environmental interests.

The key policy issue involved is whether to allow coal companies to continue dumping excess rock and dirt into streams, in violation of the Clean Water Act and the 1977 Surface Mining Control and Reclamation Act. If they are prevented from doing so, some argue, it will put the coal mines out of business. The major stakeholders in the controversy over mountaintop removal are the coal mining industry, environmentalists, area residents, state regulatory and elected office holders, and federal officials in the legislative, executive, and judicial branches. Each stakeholder has a unique perspective on the issue and a role in the controversy. Examination of these conflicting interests highlights the issue, the power and salience of various actors in this particular microcosm of state-level environmental politics, and the intermeshing of these interests in the policy process.

The coal mining industry is king in West Virginia. It staunchly supports mountaintop removal. Strip mining accounts for about one-third of the state's coal production. Spokespersons for the industry contend that an end to mountain top removal, or "strip mining on steroids," as some call it, would destroy the state's fragile economy and end all coal mining in the state.[1] The United Mine Workers of America are allied with the industry in opposing any efforts to curb the removal practice and the resulting layoffs of miners.[2] Mining proponents argue that this method allows the state's coal mining industry to remain competitive with cheaper coal imports from other parts of the United States and abroad. Further, they dispute the extent of environmental damage attributable to this practice, pointing to their successful efforts to re-contour and reseed mountaintop project sites into rolling slopes with grass and shrubs.

Environmentalists seek to stop mountaintop removal to avoid further damage to the Appalachian ecosystem. Local as well as state and national environmental interest groups have joined the fray. They are concerned that destruction of streams adversely affects fish migration and the ecosystem. Further, they feel that replacement of thousands of acres of hardwood forest with grasslands has destroyed the original contours of the land. Mountain peaks are being demolished and whole communities are getting bulldozed. Groups such as the West Virginia Highlands Conservancy has helped residents bring their cases to the federal courts.

Area residents who live near the mountaintop removal sites are unhappy with the results of using this controversial technology, and they would like to see it stopped. They claim that their houses shake, that doors and windows have been damaged, that dust is a problem, and that sheetrock and drywall are falling down. Health problems from contaminated wells and groundwater, as well as accumulated dust, are also concerns. Some residents have sold their homes, often at a reduced price, and relocated elsewhere. Others have refused to leave, and a few have stayed to fight the coal industry.[3]

The West Virginia Division of Environmental Protection (DEP), along with the state and federal government, are targets of critics. Environmentalists contend that the DEP has not met its statutory obligations to regulate the coal mines and that it has been too lax in enforcing the law and has permitted the coal industry to illegally dump rubble into streams, valleys, and hollows. Critics claim that the DEP is controlled by the coal industry, pointing out that the directors often come from the industry. The DEP claims it is doing its job—enforcing the law—and denies destroying the streams.

State elected officials, many of whom depend on sizable contributions from the coal industry to finance their campaigns, overwhelmingly support the position of the coal industry, favoring mountaintop removal. A few statewide elected officials are sympathetic to the position of the homeowners living in the mining areas and to environmentalists' concern for the Appalachian ecosystem. However, they are the exception. Former governor Cecil Underwood, previously a coal company executive, sided with the coal industry and viewed efforts to stop mountaintop removal as "effectively closing coal mining," leading to widespread job loss and loss of tax revenues. Both the state legislature and then-Governor Underwood incurred the wrath of state regulators as well as the U.S. Environmental Protection Agency when they supported a law favored by the U.S. coal industry relaxing the rules for replacing streams destroyed by mining.

Elected officials, including the late Senator Robert Byrd and other members of West Virginia's congressional delegation, also sided with the coal industry. They lobbied Congress to undo actions halting mountaintop removal, but to no avail. Senator Byrd, then powerful chair of the Senate Appropriations Committee, sought to permit the continued dumping of mine wastes in streams and valleys. He and his colleagues justified this position by arguing that dumping was necessary to protect the state's economy and miners' jobs. Their action came in the form of a rider attached to an FY2000 Senate appropriations bill. The Clinton administration initially supported this rider to accommodate Senator Byrd, but subsequently reversed itself under pressure from twenty national environmental groups to veto appropriations bills that contain anti-environmental provisions.[4]

Various federal executive agencies have been involved in the issue. At one point, the EPA threatened to take over from the DEP the regulation of West Virginia coal mines.[5] The Office of Surface Mining Reclamation and Enforcement (OSMRE) is the federal agency overseeing surface mines. Over the three decades of its existence its role has changed, from the principal regulator in most states to the overseer of state government regulatory efforts. This devolution of responsibility from the federal government to the states was necessitated by congressional budget cuts and downsizing of the OSMRE. As a result, solo inspections by OSMRE in West Virginia have stopped, and the number of joint federal/state investigations have diminished as well.[6] The EPA, OSMRE, the U.S. Army Corps of Engineers, and the West Virginia DEP entered into a memorandum of understanding specifying the conditions under which valley fills may be constructed, in addition to requirements for compliance with the Clean Water Act.

The courts have been key actors on this issue, with rulings concluding that dumping mining waste from mountaintop removal into valley streams was a violation of the Clean Air Act and OSMRE regulations. This decision stopped West Virginia's DEP from granting permits allowing such dumping in streams that flow for half a year. It was this ruling that also prompted the coal companies to close mines and lay off workers. It also led to the vigorous opposition of then-governor Underwood and the West Virginia congressional delegation. The delegation claimed the judge's interpretation of federal law was out of sync with congressional intent. The local media weighed in on the issue, with the *Charleston Gazette* supporting the court ruling in an editorial saying the "decapitation method" makes West Virginia an international example of industrial ravages.[7] The issue received national and international attention, with the appearance of numerous newspaper articles.[8] It was also the subject of a nationally televised CBS *60 Minutes* broadcast, which was sympathetic to both environmentalists and West Virginia homeowners affected by these mining methods.[9]

The court ruling was subsequently stayed. It prevented the coal industry from dumping rubble within one hundred feet of streams with year-round or half-year flows. The West Virginia delegation's attempt to pass a legislative rider overturning the court's decision passed in the Senate by a vote of 56–33, but it was not taken up prior to adjournment of the House of Representatives. Then, in the wake of the stay of the district court's ruling, the West Virginia State DEP rescinded an order stopping the activities of valley fills downstream and halting new valley fill permits. This dispute is ongoing and not likely to be resolved in the near term. It is unclear whether the final resolution will result from action by the federal courts or Congress. In 2008, just before the end of the George W. Bush administration, a coal mining rule was issued that would allow for mountaintop removal mining. A federal court decision then vacated the 2008 rule based on what Republicans claimed were "very narrow technical grounds," but Democrats insisted the vacated rule posed a risk to the environment by allowing mountaintop removal.[10] In 2009, the Obama administration criticized the Bush-era rule and indicated its intent to rewrite coal mining regulations, specifically the Stream Buffer Zone Rule regulating coal production.[11] In 2014, the Republican-controlled House passed a bill in a 228 to 192 vote to stop the Obama administration's action to rewrite coal mining regulations and require them to use instead the 2008 rule developed during the Bush administration.[12] The Preventing Government Waste and Protecting Coal Mining Jobs in America Act, H.R. 2824, would mandate not only stopping efforts to rewrite the rules, but also require the administration to assess the impact of the 2008 rule for five years before proposing new changes.

What Republicans characterized as Democrats' "war on coal" that would cost thousands of jobs and harm economic growth in twenty-two states, Democrats saw as an attempt to adequately address the negative community, environmental, and health impacts of strip mining.[13] Despite the Republican victory in the House, the Senate is not likely to consider the bill, mindful of a veto threat from the White House. Further action from the Obama White House on the issue is unlikely. This case serves as an instructive example of environmental politics at the subnational level. It illustrates the extensive range of interests affected, high stakes, conflicting pressures, players seeking to prevail in different institutional settings and at different levels of government, and how tactics change as the issue unfolds. The West Virginia case reveals the interrelation between state-level interests and national interests as well as the interconnection between social, economic, political, technological, and ecological concerns. Controversial state environmental issues like this one cannot be resolved in isolation from conditions in the broader environment and from public officials at higher levels of government.

Notes

1. Ken Ward Jr., "A Controversial Ruling Pits Miners against Environmentalists in West Virginia," *In These Times*, December 12, 1999.

2. Martha Bryson Hodel, "Union President Sees Mining 'Under Assault as Never Before,'" *The Courier-Journal*, April 2, 1999; http://www.mountaintopmining.com/newsarchive/union.htm. Accessed March 6, 2000.

3. CBS News, "King Coal; Debate over Whether a Lawsuit against a Form of Coal Mining Has Devastated the West Virginia Mining Industry as Much as Coal Companies Contend," *60 Minutes*, Season 32, Episode 24, February 27, 2000; 2000, http://web.lexis-nexis.com/universe/printdoc. Accessed February 29.

4. Peter Slavin, "A Rumble in the Hills: ARCADIA: Environmentalists Are Taking Their Protest about Open-cast Mining in West Virginia to the Top," *Financial Times* London edition, November 20, 1999.

5. Martha Bryson Hodel, "W. Virginians Resist Mountaintop-Removal Mines EPA Also Objects to State's Lax Rules as Companies Take Coal, Fill Valleys," *Milwaukee Journal Sentinel*, October 11, 1998.

6. Joby Warwick, "'Mountaintop Removal' Shakes Coal State; Cost of Prosperity Hits Close to Home," *The Washington Post*, April 31, 1998.

7. Mark Tran, "Ex-Coal Worker with Mountain to Climb to Save Valley," *The Guardian* London edition, August 14, 1998.

8. Ken Ward Jr. "A Controversial Ruling"; Ken Ward Jr., "Haden Suspends Mining Rule," Charleston Gazette Online, October 30, 1999; http://charlestongazette.wv.newsmemory.com/. Accessed September 22, 2014; Tom Kenworthy and Juliet Eilperin, "White House Backs W. Va on Mine Dumping; Conservationists Say Action Undermines Vetoes, Conflicts with Environmental Stance," *Washington Post*, October 30, 1999; Mark Tran, "Ex-Coal Worker with Mountain to Climb"; Hodel, "W. Virginians Resist"; and Slavin, "A Rumble in the Hills."

9. CBS News, "King Coal."

10. Pete Kasperowicz, "House Votes to Stop Obama's New Coal Mining Rules," *The Hill: Floor Action Blogs*, March 25, 2014; http://thehill.com/blogs/floor-action/votes/201688-house-votes-to-stop-obamas-new-coal-mining-rules. Accessed September 22, 2014.

11. Pete Kasperowicz, "GOP Issues Subpoena over Obama Mining Rules," *The Hill: Floor Action Blogs*, March 25, 2014; http://thehill.com/blogs/floor-action/energy-environment/201658-gop-subpoenas-official-over-obama-mining-rules. Accessed September 22, 2014.

12. Kasperowicz, "House Votes."

13. Ibid.

Case Study

The Dynamics of Implementation and Enforcement II: Climate Change and Sea Level Rise in South Florida

Globally, sea levels have been rising since the nineteenth century as a consequence of increasing global temperatures that have led to oceans warming, water expanding, and Arctic ice disappearing. Absent stronger controls on greenhouse gases, the United Nations Intergovernmental Panel on Climate Change projects that by the end of this century global sea level could rise as much as three feet; other reports do not rule out a rise of six feet or more.[1] It is estimated that land on which approximately 3.7 million Americans currently live could be inundated were sea level to rise by less than four feet.[2]

There is no state in the country more vulnerable to the rise in sea levels than low-lying Florida with much of its 1,197-mile coastline just a few feet above the current sea level.[3] U.S. Senator Bill Nelson has called Florida "ground zero for sea-level rise."[4] The threat to the state is evident when focusing on the 2,120 square miles of land that is fewer than three feet above high tide: there are 2,555 miles of roads, public schools, and 966 sites that the EPA has identified as hazardous waste dumps and sewage plants. At the same level, property worth $156 billion, and three hundred thousand homes, would be affected.[5] Clearly, the economic impacts of sea rise are substantial.

Protecting vulnerable shorelines from rising sea levels is a high priority, especially for South Florida, which sits at four feet above high tide; a six-foot sea level rise would put most of Miami-Dade County below sea level.[6] Indeed, Miami has been identified by the National Climate Assessment report on global warming as one of the cities most vulnerable to severe damage as a consequence of rising sea levels. Further, the Organization for Economic Cooperation and Development pinpoints Miami as the world's coastal city most threatened by sea level rise.[7] This vulnerability could result in billions or even trillions of dollars of damage, according to county government estimates, because the city's foundation of porous limestone could be soaked by rising seas, threatening fresh water and infrastructure. Unfortunately, many of those who live in high-risk, low-lying areas are disproportionately low-income people lacking the money to adequately prepare for sea level rise.[8]

Local officials in South Florida are well aware of the threat. Echoing President Obama's State of the Union address statement that "debate is settled" and "climate change is a fact," Miami Beach mayor Philip Levine says, "We are past the point of debating the existence of climate change and are now focusing on adapting to current and future threats,"[9] and the Beach's city manager has called for more "aggressive assumptions, about the effects of climate change."[10] Climate experts agree that sea levels are rising and will continue to do so, but South Florida climate scientists disagree on how much. When asked recently to estimate how high sea levels will have risen by the year 2100, the responses of local climate scholars ranged from .3 to twenty feet. While the extent of the future rise is unclear, the need for addressing the threat is not.

There is abundant evidence that national, state, and local government officials are not only deeply concerned, but also actively grappling with climate change and sea level rise. In part, the salience of the issue in South Florida is related to prior experience with flooding, inadequate storm drains, and extensive damage magnified by rising seas as well as potential devastation to beaches, homes, communities, transportation, and the economy. L. Forbes Tompkins and Christina DeConcini (2013) illustrate the effects of a three-foot rise in sea levels (almost nine hundred miles of roadway between Palm Beach and Miami-Dade county impacted), as well as a one-foot rise in Monroe County (between 65 and 71 percent of hospitals and emergency shelters would be below sea level).[11] The economic result of such developments would be in the hundreds of millions of dollars.

Already, local jurisdictions are making huge investments in plans to renovate their cities' storm water management systems (Miami has dedicated $200 million; and Miami Beach approved a $200 million spending project to improve the city's drainage system).[12] Miami-Dade County has a $1.5 billion plan outlined in a consent decree with the USEPA to repair its aging sewer system.[13] Key West is spending more than $4 million to install pumps and upgrade its drainage. Also, a new ordinance in Key West requires that all new buildings be raised above the previous standard by a foot and half as part of its strategy for dealing with climate change and sea level rise. While the South Florida business community is not as engaged on this issue as the government, the long-range impact could be devastating for business, real estate, and tourism.[14]

More specifically, Florida's Broward, Miami-Dade, Monroe, and Palm Beach counties have a combined population of 5.6 million—

exceeding the populations of thirty states and representing 30 percent of Florida's population—and an aggregate of $4 billion in taxable property values vulnerable to just one foot of sea level rise ($31 billion if seas rise three feet).[15] Spurred to action, leaders from this four-county area formed the Southeast Florida Regional Climate Compact in 2010. The compact was created to design and carry out mitigation and adaptation strategies for addressing climate change and sea level rise. This involves multijurisdictional collaboration to pursue joint policies for influencing climate-related legislation and to seek state and federal funding. In recent years, annual summits have been held for scholars, leaders, and citizens concerned about climate change and sea level rise issues where the impact of storm surges is discussed and progress is reviewed on addressing climate change and charting a path for future actions.

A variety of local, regional, state, and federal agencies (NOAA, USACE, USGS, and USEPA) provide support for the compact's technical climate work groups, including creation of vulnerability assessment and greenhouse gas inventory. Their efforts have culminated in an amendment to Florida law creating the Adaptation Action Area for areas vulnerable to climate impact such as sea level rise. With participation of myriad stakeholders, they have drafted a Southeast Florida Regional Climate Change Action Plan. The activities of the compact have captured the attention of local, national, and international entities such as the White House, media, academia, federal legislators, and nonprofit organizations (e.g., National Association of Counties).[16]

One of the first agenda items for the compact was to develop a regional climate action plan. Recommendations found in the plan were developed through a collaborative process involving nearly one hundred subject matter specialists from public and private sectors, area universities, and nonprofit organizations. The report contains 110 action items related to the following seven goal areas: Sustainable Communities and Transportation Planning; Water Supply, Management, and Infrastructure; Natural Systems; Agriculture; Energy and Fuel; Risk Reduction and Emergency Management; and Outreach and Public Policy. The implementation plan is organized as a grid broken down by goal area with specific recommendations and action steps plotted, each with an associated time horizon for a short- (0–2 years) and long-term (0–5 years) planning timeframe; potential partners; funding sources; needed policies or legislation; required resources; and performance measures or milestones.

Focusing only on sea level rise and public policy, an initial challenge was to develop a unified Southeast Florida sea level rise (SLR) projection for planning purposes. Two key planning horizons are projections of three to seven inches of SLR by 2030 and nine to twenty-four inches by 2060. The plan then maps different scenarios (one-, two- and three-foot SLRs) in each of the four counties to identify potential risk areas and plan for adaptation strategies. The goal of the public policy and outreach feature of the plan is to guide and influence local, regional, state, and federal climate change–related policies and programs through collaboration and joint advocacy.

The case is instructive because it shows how high the stakes are in environmental politics and policymaking. Collaboration among key stakeholders in adapting to sea level rise is essential. It also illustrates the interconnections among the units and levels of American government. In this instance, cooperation centered on local stakeholders, with additional assistance required from state and national policymakers. Elected and appointed government officials worked with nonprofit organizations and universities to address the key issues. Business interests are impacted by rising sea levels; a challenge is to get them more involved in designing and implementing adaptation strategies to forestall potentially devastating effects for the region.

Notes

1. The NYT Editorial Board. "Climate Disruptions, Close to Home," *The New York Times*, May 7, 2014; http://www.nytimes.com/2014/05/08/opinion/climate-disruptions-close-to-home.html?hp&rref=opinion. Accessed May 8, 2014; and Here and Now, "Elevation Zero: South Florida Prepares for Rising Sea Level," *National Public Radio*, March 10, 2014; http://hereandnow.wbur.org/2014/03/10/florida-sea-level. Accessed September 22, 2014.

2. Justin Gillis and Kenneth Chang, "Scientists Warn of Rising Oceans from Polar Melt," *The New York Times*, May 12, 2014; http://www.nytimes.com/2014/05/13/science/earth/collapse-of-parts-of-west-antarctica-ice-sheet-has-begun-scientists-say.html?_r=0. Accessed May 13, 2014.

3. Nick Madigan, "South Florida Faces Ominous Prospects from Rising Waters," *The New York Times*, November 10, 2013; http://www.nytimes.com/2013/11/11/us/south-florida-faces-ominous-prospects-from-rising-waters.html?pagewanted=all&module=Search&mabReward=relbias%3Ar. Accessed September 22, 2014.

4. William E. Gibson, "Florida Communities Prepare for Rising Seas," *Sun Sentinel*, April 20, 2014; http://articles.sun-sentinel.com/2014-04-20/news/

fl-preparing-for-rising-seas-20140418_1_rising-seas-water-south-florida. Accessed September 22, 2014.

5. Madigan, "South Florida Faces."

6. The NYT Ed. Board, "Climate Disruptions"; and Here and Now, "Elevation Zero."

7. Coral Davenport, "Miami Finds Itself Ankle-Deep in Climate Change Debate," *The New York Times*, May 7, 2014; http://www.nytimes.com/2014/05/08/us/florida-finds-itself-in-the-eye-of-the-storm-on-climate-change.html?module=Search&mabReward=relbias%3Ar. Accessed September 22, 2014; and Here and Now, "Elevation Zero."

8. Wilson Sayre, "Why Sea-Level Rise Might Hurt Poor Neighborhoods More Than Coastal Areas," *WLRN, Public Station*, April 2, 2014; http://wlrn.org/post/why-sea-level-rise-might-hurt-poor-neighborhoods-more-coastal-areas. Accessed September 22, 2014.

9. Davenport, "Miami Finds."

10. Madigan, "South Florida Faces."

11. C. Forbes Tompkins and Christina DeConcini, "Frontlines of Climate Change: Florida Leaders Take Action on Sea Level Rise," *World Resources Institute, Blog*, October 14, 2013; http://www.wri.org/blog/2013/10/frontlines-climate-change-florida-leaders-take-action-sea-level-rise. Accessed April 22, 2014.

12. Tompkins and DeConcini, "Frontlines of Climate Change"; Davenport, "Miami Finds."

13. Arianna Prothero, "Building for Sea-Level Rise—Without Rules," *WLRN, Public Station*, November 15, 2013; http://wlrn.org/post/building-sea-level-rise-without-rules. Accessed September 22, 2014.

14. Greg Allen, "Key West Awash with Plans for Rising Sea Level," *National Public Radio*, November 12, 2013; http://www.npr.org/2013/11/12/241350517/key-west-awash-with-plans-for-rising-sea-level. Accessed September 22, 2014; and Madigan, "South Florida Faces."

15. Tompkins and DeConcini, "Frontlines of Climate Change."

16. Climate Compact files, "What is the Southeast Florida Regional Climate Change Compact?" *The Southeast Florida Regional Climate Change Compact*, April 2012; http://southeastfloridaclimatecompact.files.wordpress.com/2014/05/compact_summary-document.pdf. Accessed September 22, 2014; Rosina Bierbaum, Arthur Lee, Joel Smith, Maria Blair, Lynne M. Carter, F. Stuart Chapin III, Paul Fleming, Susan Ruffo, Shannon McNeeley, Missy Stults, Laura Verduzco, and Emily Seyller, "Chapter 28, Adaptation," *National Climate Assessment, U.S. Global Change Research Program*, 2014; http://nca2014.globalchange.gov/system/files_force/downloads/low/NCA3_Full_Report_28_Adaptation_LowRes.pdf? download=1. Accessed September 22, 2014.

Conclusion

This chapter has reviewed the meaning of environmental federalism, its historical context, regulatory and nonregulatory tools, the role and performance of the states, and political dynamics and collaborative relations encountered in specific environmental contexts. While scholars disagree about the definition of federalism, they agree that government at all levels plays a crucial role in environmental protection. We have shown how regulatory "sticks" in the form of command-and-control regulations and other tools are used to enforce national standards, and how states are expected to implement such directives, often while lacking financial support from Washington. Recently, nonregulatory "carrots" have been used hand in hand with more coercive regulations. Whereas decades ago policies were based on assumptions about the limited capability and commitment of state governments to effectively protect the environment, the past three decades have reversed these assumptions as states have demonstrated increased capacity and determination to achieve environmental goals. Today, relations between the federal and subnational governments in addressing environmental concerns reflect a mixture of cooperation, conflict, and strategic decision making.

The federal environmental regulatory structure was erected in the 1970s and extended in the 1980s, but the 1990s and the first fifteen years of the 2000s have witnessed the diminished enthusiasm, and in some instances active hostility, of national policymakers for new federal environmental legislation. State and local policymakers now vigorously object to new federal mandates when resources do not accompany such directives. Yet when states have the will, resources, and capacity to act, they have achieved success. But state environmental initiatives remain uneven: some states take the lead and follow best practices, others lag behind, unwilling or unable to undertake innovative action to protect the environment. This substantial variation in effort and performance among the states is also evident in the extent of trust and involvement in federal-state relationships.

The two case studies in this chapter (Mountaintop Removal and Sea Level Rise) highlight the controversial nature of environmental politics and the complex tradeoffs that often exist in the calculation of economic and environmental gains and losses. They show that political actors from all levels of government, citizens, organized interests, and the private sector have crucial stakes in the outcome of these contentious and often costly disputes. The strategic maneuvering of these key stakeholders is a political game that is shaped by legal, economic, social, technological, and ecological considerations. In West Virginia, the battle lines have pitted environmentalists against powerful coal mining interests over "mountaintop removal" strip mining. In South Florida, the environmental threat of sea level rise has pushed South Florida officials and other stakeholders to join together to forge adaptations and implementation strategies for both the short and longer term. Unlike West Virginia, partisan politics were muted in the South Florida case, where consensus exists that inaction is untenable and cooperation essential. In each case both the "top-down" influence of national government officials and the "bottom-up" clout of state and local actors were critical in the political interplay that ensued. While the future of federal-state relations in environmental policy is still unfolding, as it is in these two cases, it is clear that successful solutions to environmental problems will require joint efforts by those at the top, middle, and bottom of the federal system.

Websites

American Federalism and Environmental Politics

Council of State Governments: www.csg.org

International City/County Management Association: www.icma.org

National Association of Counties: www.NACo.org

National Conference of State Legislatures: www.ncsl.org

National League of Cities: www.nlc.org

National Governors Association: www.nga.org

State and Local Government on the Net: www.piperinfo.com/state/states.html

3

Public Opinion, Interest Groups, and the Environment

In this chapter, rather than narrowing our focus to one institution, we address two similar but distinct aspects of the public's connection to environmental politics and policy—namely, public opinion and interest groups. Citizens are linked to the policy process through public opinion polls and organized interests play an integral role in linking citizens to the policy-making process.

For the purpose of this chapter, therefore, it is important that we gauge public opinion about environmental protection because "broad public support in favor of environmental protection provides legitimacy for those working on its behalf."[1] This becomes increasingly important as American citizens are challenged by a variety of threats, including global climate change and its varied consequences (e.g., sea level rise and extreme weather events). The discussion that follows also addresses the role of interest groups that are focused on environmental politics and policy. We will examine the role, characteristics, and activities of the groups acting under the umbrella referred to as the environmental movement. In our examination of environmental organizations, our goal is to improve our understanding of the nature of interest group formation and activism as it relates to environmentalism and ecological issues.

American Public Opinion and the Environment

One of the interesting aspects of the American public's consideration of environmental issues has been the general consistency in the public's attitudes about protecting the environment. Four decades ago, Anthony Downs presented an "issue attention" model in order to explain the evolution of public policy problems.[2] The model developed by Downs suggested that all public policy problems emerge, mature, and decline in a five-stage process. Namely, (1) the pre-problem stage; (2) the alarmed discovery and public desire for the government to do something about it stage; (3) the realization of the costs of progress stage; (4) the decline in public interest in the problem stage; and finally (5) the post-problem stage in which new problems arise, displacing the original issue. Nevertheless, the American public's interest in environmental protection remains quite strong and therefore does not necessarily end with the final, post-problem, stage. The American public has remained notably concerned about the quality of the environment.

Public Opinion and National Goals: Environmental Protection or Economic Growth?

The first step in constructing a portrait of American opinion about the environment is to examine American citizens and determine their preference between two important, opposing goals—namely, environmental protection and economic growth. These two choices have been presented to Americans in Gallup Polls over several decades, providing insight into citizen preferences regarding these two important national issues. Gallup Poll data illustrates the following pattern during the period 1984–2014.[3] In 1984, Americans preferred environmental protection 61% to 28%. Ten years later (1994), American public opinion remained consistent, as six out of ten Americans preferred protecting the environment, compared to 30% who favored economic growth. A decade later (2004), although public support for environmental protection had dropped to 49%, it still surpassed the 44% who favored economic growth. Ten years later (2014), while 50% of Americans preferred protecting the environment, support for economic growth stood at 41%. In short, although American citizens' preference for environmental protection has dropped (due, in part, to the Great Recession), it remains an important goal for the country.

Another way to assess the comparative difference among Americans on these two important goals is to examine preferences based on several sociodemographic characteristics—namely, party preference, gender, and age. Gallup Poll data illustrate the impact of partisanship during the period 1998–2014.[4] In 1998, 56% of the Republicans surveyed preferred environmental protection and 35% favored economic growth while 73% of the Democrats favored protecting the environment compared to 21% preferring economic growth. A decade later (2008), while one-third of Republicans favored protecting the environment, almost six out of ten (59%) favored economic growth. At the same time, 59% of Democrats supported environmental protection and 32% supported economic growth. In 2014, six out of ten (59%) Republicans supported economic growth and one-third supported environmental protection while two-thirds of the Democrats favored environmental protection and 27% supported economic growth. All in all, Democrats have been very consistent in their support for environmental protection over economic growth and Republicans have been similarly supportive of economic growth over environmental protection.

How do Americans view these two goals in terms of gender and age? During the brief period March 2013 to March 2014, we find an interesting profile of Americans in terms of these two variables regarding environmental protection and economic growth.[5] First, women, by a small margin, supported environmental protection over economic growth compared to their male counterparts. Second, Gallup Poll data indicated that younger people were more likely to support environmental protection than older Americans. In fact, support among young people increased from 49% in 2013 to 60% in 2014. Second, both young men and women were more likely than their older counterparts to support environmental protection. At the same time, it is clear that as one grows older, support for environmental protection tends to decline for all groups and support for economic growth increases.

Public Opinion and National Issues: Where Is the Environment?

Based upon data produced in the March 2014 Gallup Poll, a portrait of the American public can be established in terms of the extent to which they worry a "great deal" about fourteen national issues.[6] The percentage of Americans worrying a "great deal" can be organized as follows: upper tier (economy 59%, federal spending/budget 58%, healthcare 57%), middle tier (unemployment 49%, size/power of the federal government 48%, Social Security 46%, hunger/homeless 43%) and bottom tier (crime/violence 39%, terrorist attack 39%, affordable energy 37%, drug use 34%, illegal immigration 33%, quality of the environment 31%, climate change 24%, race relations 17%). While the American public's concerns in 2014 are relatively similar to those found in earlier polls, what appears evident is that "environmental quality" remains at the bottom of the list of fourteen major issues.

One way to explain this "low ranking" of the environment in relation to other national issues is to examine partisan similarities or differences. An examination of the fourteen national issues shows that Democrats are more worried about protecting the environment compared to Republicans.

Democrat and Republican concerns about the environment reflect a partisan gap of twenty-nine points. In short, Americans' low concern about environmental quality is tied to partisan differences in general and reduced concern by Republicans in particular. This is reflected in a January 2014 poll where Democrats and Republicans were asked to list their Top 10 Priority Issues for Congress and the president.[7] Democrats listed the environment in the Top 10 issue concerns, but Republicans failed to do so.

Public Opinion and Environmental Issues

What has been the public's level of concern about several major environmental issues during the period 2000–2014?[8] We can say with certainty that Americans remain very concerned about issues related to water resources. Drinking water, water pollution, and toxic waste threats to land and water remain at the top of the list. Furthermore, a large minority (between four and five out of ten) of Americans continue to worry "a great deal" about two other issues—air pollution and biological diversity. At the same time, only one-third of Americans worry "a great deal" about global warming. At the beginning of the twenty-first century, a majority of Americans worried "a great deal" about five of the seven issues. About a decade later (in 2009), and again in 2014, only three of the seven issues were found to lead Americans to worry "a great deal" about them. Having said this, we would argue that Americans may be described as less likely to worry "a great deal" about these issues if they feel that threats to the environment have declined.

Case Study

Climate Change and Public Opinion

Ninety-seven percent of climate scientists agree that global climate change is occurring and that human activities are a major contributor to this climatic phenomenon. The scientific community has provided American citizens with the purported consequences of global warming and climate change on American society. For instance, scientists have informed us that a variety of environmental, economic, social, political and national security problems are on the horizon while some outcomes are already being seen. Given this background, as Sussman and Daynes have stated, "while organized interests have played an important role regarding the climate change issue, it is also necessary to examine and assess public opinion among American citizens in order to ascertain the extent to which they are knowledgeable about the issue and what they think should be done about it."[1]

Two interesting and problematic concerns about the climate change issue are the complexity of the issue and its relatively recent place on the public agenda in the United States and other countries. Compared with air and water pollution issues and/or hazardous waste concerns that have a state or local impact, climate change is global in nature and it is difficult for average citizens to grasp the complexity of the issue. In short, research findings of the scientific community have not been easily integrated into American discourse and understanding. Having said that, let us provide a portrait of American opinion on the subject, focusing on what Americans believe to be the causes of climate change.

Table 1 shows American opinion about the causes of climate change over the last decade. When asked to indicate which factors—natural conditions or human activities—are primarily responsible for climate change, Americans have been fairly consistent in suggesting that human activities (burning of fossil fuels) account for the warming of the planet and a changing climate. In short, despite some fluctuation in the findings, it is clear that in each year, a majority of Americans focus on human activities rather than natural causes as the source of the problem.

Table 1. U.S. Citizens' Views About the Causes of Climate Change

	2003	2006	2010	2012	2014
Human Activities	61%	58%	50%	53%	57%
Natural Causes	33%	36%	46%	41%	40%

Source: Adapted from Lydia Saad, "A Steady 57% in the U.S. Blame Humans for Global Warming," March 18, 2014 at www.gallup.org/poll/167972-steady-blame-humans-global-warming-.aspx.

One of the fundamental factors in American politics that helps explain citizens' attitudes about policy issues is political partisanship. To what extent might political partisanship influence one's opinion on climate change? Table 2 presents research findings that compare Democrats, Independents, Republicans, and Tea Party identifiers and their views about the causes of climate change. As we can see, Democrats more than Independents, Independents more than Republicans, and Republicans more than Tea Party identifiers believe that changes in the climate are due to human activities.

Table 2. The Impact of Partisanship on U.S. Citizens' Views About the Causes of Climate Change, 2014

Human Activities	*Natural Causes*
Democrats 79%	21%
Independents 50%	50%
Republicans 41%	59%
Tea Party* 19%	81%

Note: Tea Party percentages are from 2011 adapted from A. Leiserowitz, E. Maibach, C. Roser-Renouf, and J. D. Hmielowski, "Political and Global Warming: Democrats, Republicans, Independents, and the Tea Party" (New Haven: Yale Project on Climate Change Communication, 2011), at http://environment.yale.edu/climate/files/PoliticsGlobalWarming2011.pdf.

Source: Adapted from Lydia Saad, "A Steady 57% in the U.S. Blame Humans for Global Warming," March 18, 2014 at www.gallup.org/poll/167972-steady-blame-humans-global-warming-.aspx.

Polling data provide us with a better understanding of the political orientations of American citizens when it comes to climate change. However, moving from opinions to policymaking is problematic. Polarization among members of the U.S. Congress is quite evident on the issue of climate change. This suggests that we will continue to see political conflict over how to reduce greenhouse gas emissions in the years to come due to partisanship, ideological differences, and opposition by economic interests opposed to making positive changes.

Note

1. Glen Sussman and Byron W. Daynes, *U.S. Politics and Climate Change* (Boulder: Lynne Rienner, 2013), 143.

A Clean Energy Agenda

Against the background of the preceding sections, a related issue concerns U.S. energy policy. To what degree do Americans have a preference to continue reliance on fossil fuels compared to pursuing a clean energy agenda? A clean energy agenda would emphasize energy conservation and increased use of alternative sources of energy.[9] On the question of energy conservation and conventional energy production, American public opinion has been quite consistent. Over the last fifteen years, almost six out of ten Americans have favored energy conservation over conventional energy production. Only one-third of Americans prefer continued reliance on conventional sources of energy—oil, coal, natural gas. Moreover, preference for a clean energy agenda is found among younger citizens who will, over time, replace older citizens and their preference for maintaining the energy status quo.

In addition to generational differences regarding a clean energy agenda, partisanship continues to have an impact on Americans' energy choices. Several observations can be made based on the data in Table 3.1. First, although a majority of citizens regardless of partisanship support increased emphasis on alternative sources of energy, a conspicuous gap separating Democrats and Republicans is evident. Second, Republicans are much more likely to support reliance on conventional energy sources compared to their Democratic and Independent counterparts. Having said this, a decidedly large gap separates Republicans from Democrats. Third, among conventional energy sources, natural gas is the source of choice for all three groups. Access to natural gas results from a process called fracking that results in environmental negative externalities. More data are needed to better ascertain the American public's views about natural gas and the fracking process. Finally, where almost one-half of Republicans support continued emphasis on nuclear power, only one-third of Democrats do so. One advantage of nuclear power is that it does not produce greenhouse gases that contribute

Table 3.1. Support for Conventional and Alternative Sources of Energy by Party

	Democrats	Republicans	(Gap)*
Conventional Energy Sources			
Coal	21%	51%	−30
Natural gas	59	78	−19
Nuclear Power	30	49	−19
Oil	29	71	−42
Alternative Energy Sources			
Solar power	87%	68%	+19
Wind	83	59	+24

Note: An (*) represents the gap between Democratic and Republican partisans. A (+) indicates more support for alternative energy sources while a (-) indicates more support for conventional energy sources.

Source: Adapted from Dennis Jacobe, "Americans Want More Emphasis on Solar, Wind, Natural Gas," March 27, 2013; www.gallup.com/poll/161519/americans-emphasis-solar-wind-natural-gas.aspx?version. Accessed April 20, 2014.

to global warming and climate change. However, serious questions remain about the security and integrity of nuclear power plants and how and where to store nuclear waste.

Governmental Performance and Environmental Protection

Over the past several decades, although there has been some fluctuation in the political orientations of citizens with regard to the issue of protecting the environment, Americans overall have shown strong concern for securing a healthy environment. We will point this out as we question concerns for governmental performance.

One way to describe the public's political orientation toward the level of governmental progress in the environmental policy arena is by citing the extent to which citizens view the government as doing "too little," "too much," or "about the right amount" to promote environmental protection. Gallup Poll data inform us that two decades ago 68% of Americans indicated that the U.S. government was doing "too little" to protect the environment.[10] A decade later, as the country was beginning a new century 51% believed that the federal government was doing "too little"—a 17 percent decline from 1992. Since then, how has the federal government fared in the eyes of the American public in terms of making progress in support of environmental protection? Gallup Poll data show us that in 2013, 47% of Americans believed that the federal government was doing "too little" in support of a healthy environment, down from 51% who felt this way just a few years earlier.[11] What are we to make of these findings during the period 1992–2013? We must conclude that there was a decline of 21% among those who felt government was doing "too little." This suggests that, overall, the American public believes that the federal government has been playing a positive role in protecting the environment. On the other hand, almost half (47%) of Americans still hold the view that the government is doing "too little" suggesting that much more work remains to be done by the federal government.

The preceding discussion provides a portrait of the American public in terms of the extent to which it believes that the federal government has made progress regarding the environment as an important policy issue. Given these results, what types of policies do Americans prefer that government use in protecting the environment? In March 2014, nine proposals were addressed by the Gallup poll.[12] First, more than six out of ten Americans support strong enforcement of environmental regulations. Examples include higher emissions and pollution standards for business and industry, mandatory controls on carbon dioxide emissions and other greenhouse gases, and higher emissions standards for automobiles. Second, the public supports directing more tax dollars toward support for the environment. Additional examples include spending more money on alternative sources of energy (solar and wind) and spending more money to develop alternative fuels for automobiles. The one anomaly among these nine proposals was the public's support for oil exploration on public lands (a 58% majority). The one option that failed to receive a majority of support from the public (47%) was greater employment of nuclear power.

Given this background, it is important to close this section by focusing on two specific environmental preferences held by Americans regardless of partisanship—environment and jobs, and government energy subsidies—preferences that may well provide a look into the future of American politics and the environment. The first question deals with the effect of increased environmental regulation on jobs. More often than not, Congressional Republicans have argued that increasing regulations on the energy industry will reduce the number of jobs for Americans. This has been used to justify opposition to the environmental regulatory process. How do American citizens feel about this issue? In May 2012, poll data from a study by Yale University and George Mason University's climate change program indicated that almost six out of ten (58%) Americans believed that job growth was positively related to protecting the environment.[13] The public and congressional Republicans were diametrically opposed on this issue. The second question addressed the issue of providing federal subsidies for the energy industry. Poll data compiled by the nonpartisan Civil Society Institute showed that a majority of Americans across party lines (including Tea Party identifiers): (1) opposed subsidies for the fossil fuel and nuclear power industries, and (2) wanted federal subsidies shifted to support alternative, renewable sources of energy such as solar, wind, and technologies geared toward energy efficiency.[14]

Interest Groups and the Environment

The environment is one among many public policy issues that poses challenges for governmental policymakers and organized interests. Although the environment has been a salient aspect of modern American politics, when and to what extent have citizens organized into formal groups in order to express their concerns about environmental protection?

Although environmental interest groups are very active in the American political setting, they are not necessarily uniform in their tactics, strategies, and goals. Environmental groups have shared interests regarding the overall objective of protecting the environment, but differ in many ways. For example, large organizations such as the National Wildlife Federation have memberships and resources that dwarf smaller groups like Friends of the Earth. Where some groups have broad concerns, others are much narrower in their focus. Some groups are primarily mainstream and legitimate in their political activism while the actions of other groups might be considered extreme or radical. Some environmental organizations are characterized by formal, legitimate practices while other groups exhibit what are referred to as unconventional, direct action techniques.

Personal Profile

Robert F. Kennedy Jr.

As a member of the large, politically active Kennedy family, it is not surprising to find Robert F. Kennedy Jr. involved in public life. However, rather than seeking public office following in the footprints of his father, Robert, and his uncle John, Robert Jr. decided to pursue an alternative path. As an environmental activist, he has employed a variety of tools in his effort to protect the environment in general and drinking water and waterways in particular. On the one hand, for instance, he is the founder of and chief prosecuting attorney for the organization, Riverkeeper. He also serves as chair of the Waterkeeper Alliance and is the senior attorney for the well-known group Natural Resources Defense Council. On the other hand, he has also been involved in what is referred to as unconventional protest behavior. For example, in February 2013, he, along with the head of the Sierra Club and several other celebrities and activists, was arrested for blocking access to the White House during a protest demonstration.[1] The purpose of the protest was to express opposition to the Keystone XL oil pipeline and encourage President Obama to oppose its construction.

Kennedy has used his position in public life to support and oppose specific environmental issues. He has been an ardent supporter of wind power and a harsh critic of government subsidies to the oil and coal industries. On one hand, he has been praised for delivering "a passionate defense of the environment and how its continued neglect affects the future of the planet and the health of future generations."[2] On the other hand, he can be critical in the public sphere, as he was when he "referred to several media personalities (e.g., Glenn Beck, John Stossel, Sean Hannity and Rush Limbaugh) as 'flat-Earthers' and 'traitors.'"[3]

Kennedy has also been involved in the global climate change issue, and has used his position to criticize the role played by the fossil fuel industry. For example, in early 2013, during an interview about climate change with author and founder of the group 350.org Bill McKibben, Kennedy argued that President Obama needs to do much more about climate change than he had during his first term in office.[4] Eighteen months later, in response to the June 2014 proposal by President Obama to reduce carbon emissions, Kennedy criticized the Cato Institute (financially supported by the Koch brothers) for taking the position that Obama's proposal would have only a minimal impact and suggesting that new carbon regulations should be opposed because they would be harmful to the economy.[5]

In advance of a talk in Cleveland, Ohio, in 2012, *EcoWatch*, an environmental news website referred to him in the following way:

> Kennedy's reputation as a resolute defender of the environment stems from a litany of successful legal actions. He was named one of *Time* magazine's "Heroes of the Planet" for leading the fight to restore the Hudson Bay and *Rolling Stone* magazine's "100 Agents of Change."[6]

He has authored several books, including *New York Times* best-seller *Crimes against Nature* (2004), which called into question environmental policies in the U.S. during the first term of George W. Bush.

Notes

1. "Daryl Hannah, Robert F. Kennedy, Jr. Arrested at White House Protest against Keystone Oil Pipeline," *New York Daily News*, February 13, 2013; http://www.nydailynews.com/news/politics/daryl-hannah-rfk-jr-arrested-keystone-pipeline.

2. "Author and Environmental Activist Robert F. Kennedy, Jr. Comes to Cleveland for Town Hall Series." October 8, 2012; http://ecowatch.com/2012/10/08/robert-kennedy-jr-comes-to-cleveland/.

3. RFK Jr., "Live earth," July 7, 2007, *You Tube*; http://www.youtube.com/watch/v=Heku90TL7ysg.

4. Transcript of the *Ring of Fire* interview of Bill McKibbon conducted by Robert F. Kennedy Jr., "Time to Act on Climate Change," January 9, 2013; http://www.desmogblog.com/print/6783/.

5. Brandon Baker, "Robert F. Kennedy Jr. Praises Obama's Carbon Rules, Blasts Koch brothers on 'The Ed Show'"; http://ecowatch.com/2014/06/03/robert-f-kennedy-jr-obama-carbon/.

6. "Author and Environmental Activist Robert F. Kennedy Jr., Comes to Cleveland for Town Hall Series," October 8, 2012; http://ecowatch.com/2012/10/08/robert-kennedy-jr-comes-t0-cleveland-for-town-hall-meeting.

Membership and Strategy of Environmental Groups

We live in an age when interest groups have proliferated in American politics. Environmental interest groups are but one subset of a variety of interests, each seeking to influence public policy in the nation's capital as well as in the fifty states and at the local level. While most interest groups engage in what one might call legitimate political action (e.g., lobbying), other groups prefer alternative participatory avenues (e.g., direct action) in order to publicize their cause and influence policy. In an analysis of national interest groups, Christopher Bosso suggested that environmental organizations can be classified into five basic types: (1) large groups that focus on many issues, (2) small groups with a more narrowly defined focus, (3) nonpartisanship groups that emphasize education and research, (4) groups that promote solutions based upon legal and/or scientific grounds, and (5) groups that protect land and ecosystems through their purchase and eventual preservation.[15] What these groups have in common is their emphasis on legitimate, conventional political expression. Conventional political action includes a multitude of avenues for political action as organized interests seek to shape public policy. These methods of political action include, but are not limited to, traditional activities such as coalition formation, legislative testimony, grassroots organizing, litigation, use of news media, polling, and modern communication techniques including the Internet.

Research shows that alternative means of political expression are also put into practice by segments of the public, both in the United States and in other countries.[16] These types of actions are considered "unconventional" because they are considered outside the mainstream of typical political activism and include direct action techniques that might be considered radical or in some cases unlawful. In contrasting "ordinary" and "extraordinary" politics, Charles Euchner has argued that, while the former includes a "system of competition" but fails to provide "an effective means to challenge dominant values," the latter "aims to force the political establishment to address issues that it would rather ignore."[17] Nonetheless, members of certain environmental groups view aggressive, direct action modes of political behavior (e.g., illegal demonstrations, blocking traffic) as appropriate means by which to engage in environmental protection.

Disagreements are evident in the debate over the extent to which the public should have input into public policymaking.[18] However, the proliferation of environmental groups and the increase in their membership over the years is a clear indication of the level of public concern about the quality of the environment.[19] Moreover, to what extent do citizens contribute financially to support organized interests? It is not surprising to find that when citizens feel that the environment is receiving due consideration by the government, the level of financial contributions to organized interests levels off. Research also shows that while donations to environmental groups decline during times of unemployment, partisanship still matters to the public. Thus, when the Republican party has had control of the White House, financial contributions to environmental groups have increased.[20]

In order to provide a brief portrait of environmental groups over the last eight decades, we focus on the characteristics of a select group of national environmental groups. The earliest groups date back to the end of the nineteenth century and extend into the middle of the twentieth, and include the Sierra Club (1892), National Audubon Society (1905), Wilderness Society (1935), National Wildlife Federation (1936), and Defenders of Wildlife (1947). They all began with a common focus of concern for protecting public lands and wildlife. Examples of groups established since the emergence of the environmental movement in the 1960s include the Environmental Defense Fund (1967), Friends of the Earth (1969), and the Natural Resources Defense Council (1970). These organizations have focused on threats posed to the environment by pollution, conservation of natural resources, and the contemporary "third stage" issues of biodiversity and climate change.

Having said this, it is important to point out that many of these environmental organizations are no longer limited to one or two issues, but have evolved into multi-issue groups. For example, although the National Wildlife Federation began as a movement concerned with public lands and wildlife, it has become one of the largest environmental groups engaged in a vast array of environmental issues including air and water quality, biodiversity, stratospheric ozone depletion, and global climate change.

Memberships and budgets are critically important to organized interests. As the number of environmental groups has expanded over the years, so have the memberships of these groups. During the period 1975–2014, the membership base of environmental groups expanded tremendously overall, although it fluctuated within that time period.[21] Moreover, the increase in numbers was greater for some groups than for others. For instance, large increases can be found among the following: the Sierra Club increased to 2.4 million members from 550,000 members; the Natural Resources Defense Council saw its numbers grow from 120,000 to 1.4 million; and the Environmental Defense Fund reached a membership of 750,000, up from 300,000. In contrast, while the National Wildlife Federation grew in numbers, the rate of growth slowed for this organization, only increasing to 4.4

million members from 4 million. At the same time, the National Audubon Society saw its membership decrease from 550,000 members to 403,000.

In addition to membership and resources, environmental groups, like other organized interests, require effective leadership and coalition building. As environmental groups have matured, new leaders have pursued strategies that some consider less aggressive, albeit still effective, and at the same time they have broadened their approach by pursuing coalition formation with other sectors of society as a way to promote environmentalism.[22] Environmental groups have also aligned with other interests, including business and industry, in order to avoid the kind of conflict that existed in the past. "Green" business is an example of this new orientation. For instance, a variety of companies and other organizations, including Intel, Microsoft, Google, Staples, Apple, the U.S. Environmental Protection Agency, Best Buy, FedEx, Starbucks, and Georgetown University have increasingly been utilizing green power energy sources (e.g., solar power, wind power, biomass).[23]

Notwithstanding these examples, environmental groups remain concerned about the role played by business and industry. For example, although Exxon Corporation publicized its decision to contribute several million dollars toward the Save the Tiger Fund in the 1990s, the contribution was only a fraction of the tremendous profits made by the oil company that uses the tiger as its corporate logo. Exxon also received criticism for its role in the *Exxon Valdez* oil spill in 1989 and has been one of the more vocal opponents regarding the need to address the challenges posed by global climate change.

Against a background of positive action during the 1960s and 1970s, the environmental community was compelled to increase its attention to the White House and the Congress. Regarding the presidency, environmental groups were concerned about the environmental orientation of presidents Ronald Reagan and George W. Bush. During his second year in office, a coalition of environmental groups led by Friends of the Earth flexed its muscles by publishing an environmentalist indictment against the presidential administration of Ronald Reagan and its policies. The indictment was quite comprehensive, addressing a range of governmental shortcomings, neglect, and malfeasance in areas ranging from air and water quality to hazardous wastes to public lands to fish and wildlife. According to this document, the Reagan administration, unlike other presidential administrations, had turned away from longtime bipartisan support of environmental protection. In the words of the report:

> President Reagan has broken faith with the American people on
> environmental protection. . . . [H]e and his appointed officials

have simply refused to do the job that the laws require and that Americans expect of their government—to protect the public health from pollution and to use publicly owned resources and lands for the public good. . . . In the name of "getting the government off our backs," they are giving away our natural heritage.[24]

It is important to note that a decade prior to the Reagan administration, Richard Nixon, who shared the same party label as Reagan, stood in stark contrast to him regarding environmental policy. Where Nixon supported the establishment of the Environmental Protection Agency, signed the National Environmental Policy Act, and supported Earth Day, Reagan became a clearly identifiable enemy who could be used by the environmental community in its efforts to recruit members. Moreover, public opinion in support of environmentalism was clearly at odds with Reagan's approach to environmental policy.

Twenty years after the election of Ronald Reagan, George W. Bush moved into the White House. Rather than following the more moderate approach set forth by his father, Bush the son chose to follow the path set forth by Ronald Reagan, thus pitting himself against environmentalism by opposing the use of government authority and resources to support environmental initiatives and the regulatory process, and stood steadfast against efforts to address the global warming/climate change issue. The administration of George W. Bush also questioned the science of climate change, and actually directed members of his administration to rewrite or revise existing scientific reports that contained documentation of threats to the environment.

Although the elections of Democrats Bill Clinton in 1992 and Barack Obama in 2008 led environmentalists in each case to believe that they had an ally in the White House, they were alarmed and mobilized in response to the actions of Republican majorities in the Congress. Republicans controlled both chambers of the Congress for six of Clinton's eight years in office and in 2014 Republicans gained control of the Senate having had a majority in the House of Representatives since 2011. While some might have argued that this signaled a shift in the public's orientation toward public policy, including the environment, environmental groups, supported by public opinion polls, argued differently. For instance, Jay Hair, president of the National Wildlife Federation stated that "[a]nyone who thought this election [1994] was a mandate to undo 25 years of environmental protection had better think again."[25] The same could have been said about the midterm election of 2014.

Table 3.2. Campaign Contributions to Congressional Democrats and Republicans by Environmental Interest Groups, 1994–2014

	Total Contributions (in dollars)	To Democracts	To Republicans
1994	$1,896,977	95%	5%
2004	4,287,577	92	8
2014	83,303,349	91	9

Adapted from the Center for Responsive Politics at www.opensecrets.org/industries/totals.php? cycle=2014&ind=Q11/. Accessed May 10, 2014.

Although environmental organizations share a common commitment to environmental quality both at home and abroad, the means by which these groups pursue their goals differ considerably. One common thread that characterizes environmental groups, however, concerns campaign contributions. As Table 3.2 indicates, during the twenty-year period from 1994 to 2014, environmental organizations clearly sided with Democratic congressional candidates. In short, Democrats received the lion's share of campaign contributions distributed during electoral campaigns. All in all, campaign contributions by environmental organizations paint a picture of strong partisanship when it comes to parties and the environmental policymaking process.

Diversity of Interest Group Political Activism

Large environmental groups with mass membership and large budgets have the capacity to engage in multiple activities in their effort to influence and shape government and reach out to the public. They have done so using conventional participatory tactics. For instance, the Sierra Club, one of the oldest groups, and the Wilderness Society, founded during the Great Depression, have been engaged in efforts to protect wilderness and public lands.[26] In the course of doing so, they have been challenged by opposition interests including mining, oil, gas, and logging—interests that want to maintain or expand their own access to vital and profitable resources. In an effort to set aside public lands as wilderness areas or national parks, conservation efforts have been met with strong opposition by these interests, who argue that resources are being unnecessarily locked up. Nonetheless, these conservation organizations cite numerous successes, including the 1964 Wilderness Act, the 1980 Alaska National Interest Lands Conservation Act, and the 1996 Grand Staircase-Escalante National Monument in Utah. Taken together, millions of acres of public lands have been set aside for future generations due, in part, to the activities of these above mentioned groups as well as other environmental groups.

As the largest environmental organization in the country, the National Wildlife Federation (NWF) has been active in environmental affairs since its beginning more than eight decades ago. Started in the mid-1930s at the same time that the first North American Wildlife Conference was organized during the administration of Franklin Delano Roosevelt, the NWF became an influential force in the environmental community. Eight decades later, the NWF is involved in numerous activities including education campaigns, lobbying, network broadcasting, litigation, and product merchandising, among others.[27] The NWF has been a vigorous participant in promoting air and water quality, toxic waste cleanup, and protection of biodiversity and endangered species, and in providing solutions to global climate change.

Founded during the mid-1960s, the Environmental Defense Fund (EDF) is an example of an organization that has moved beyond focusing on public lands and wildlife to concern for second-tier issues such as hazardous and toxic waste. Pollution control and the threat posed by toxic and chemical waste have been at the center of activity for many environmental groups. The EDF was established as a result of concern raised about pesticides, and especially DDT, first brought to the public's attention by Rachel Carson in her book *Silent Spring*. Although the efforts to address the problems posed by DDT began during the Kennedy administration, DDT was not banned nationally until the end of Nixon's first term in office. The EDF stressed the importance of the public and environmental threat posed by DDT and took a novel approach, namely, using the courts to establish environmental law.[28]

While groups such as the Sierra Club, the Wilderness Society, and the National Wildlife Federation, for example, have used an array of conventional tactics to further their political agenda, other groups, including Greenpeace and Earth First!, have employed direct action techniques in support of their environmental agenda. However, where Greenpeace remains within the legitimate sphere of organized interests, Earth First! is characterized as a radical group by some policymakers and a violent organization by others. Greenpeace has members in both the United States and other countries. The organization has engaged in activities that fit neatly into its media campaigns, which publicize its efforts as well as what it considers threats to the environment or wildlife. Nonviolent, as well as direct action, techniques employed by members of Greenpeace include placing themselves in small boats between whales and whaling ships or hanging large banners from tall buildings or bridges. When Greenpeace sailed a vessel into a French nuclear test zone in the South Pacific in the early 1970s, it was challenged by a French military vessel and sustained some damage. The French military completed its nuclear test.[29] Although Greenpeace has gained national as well as international publicity through its direct action campaigns, it has expanded its repertoire of activities to include scientific research as well as working with consumer groups and business and industry in order to achieve its goals.

As a counterpart to Greenpeace, and other direct action groups, Earth First! has been characterized as a radical organization that engages in political acts, including violent protest behavior. As a consequence of its philosophical orientation, Earth First! has engaged in "ecotage," which includes, among other things, attempts to block bulldozers from gaining access to forests, and posing a direct threat to loggers who cut down trees with chainsaws, by placing metal spikes into trees to prevent timber companies from harvesting them.[30] According to one political observer of such unconventional politics:

> Unlike the mainstream national environmental organizations, which work within established political and economic frame-works, Earth First! challenges those who embrace environmental compromise and pursue environmental objectives through traditional Madisonian interest group and lobbying processes. Instead, members of Earth First! believe that the natural world must be defended through direct action, civil disobedience, and eco-sabotage.[31]

As these examples clearly show, the effort to protect the environment does not entail uniform political practices. To the contrary, environmental groups have exhibited a diverse array of tactics and strategies in their attempts to secure their environmental goals. While the vast majority of members of environmental groups belong to organizations that pursue legitimate political activism, a minority of members are willing to engage in activities characterized by nonviolent direct action or even, at times, violent methods.

Case Study

Interest Group Activism and the Place of Wolves in the Rocky Mountain Region

Wolves are a natural part of the North American ecosystem, yet they have all too often been viewed with fear due to their purported threat to cattle and livestock interests, who have long supported their removal by death or relocation. Although wolves had roamed freely in the northern Rocky Mountains, in 1926 the U.S. government sponsored an eradication program that resulted in their removal from Yellowstone National Park.[1] After nearly seven decades, the U.S. Fish and Wildlife Service (FWS) issued and the Department of the Interior adopted a wolf recovery program. Based upon the FWS environmental impact statement, "The Reintroduction of Wolves to Yellowstone National Park and Central Idaho," the stage was set for wolf recovery in 1994.[2]

The process involved introducing Canadian wolves into Yellowstone National Park and Idaho. However, although wolves were reintroduced into Yellowstone National Park in 1995, cattle and livestock interests represented by the American Farm Bureau (AFB) filed a lawsuit against the Department of the Interior to reverse the program. The AFB argued against the federal government's interference in state affairs. Moreover, the organization suggested that the interests of the wolf were being placed above those of ranchers and farmers.

In December 1997, William Downes, a Wyoming federal judge, ruled that the wolf reintroduction program was illegal and therefore the Canadian wolves were to be removed from Yellowstone and Idaho. Environmental groups stepped forward to appeal Judge Downes's decision. Included in the judicial appeals process were the National Wildlife Federation (NWF), Defenders of Wildlife, and the National Audubon Society. Moreover, the Interior Department had joined with environmental groups in opposing the AFB and the decision by Judge Downes. These groups argued that wolves, as predators, have an important role to play in maintaining the natural balance in the Rocky Mountain ecosystem and did not pose a threat to livestock interests.

Oral arguments were heard by the Tenth Circuit Court of Appeals. After a long court battle in which environmental groups opposed livestock interests, in early 2000 the court ruled, in effect, that wolf reintroduction in the Rocky Mountain region was acceptable. In response to the court's decision, Mark Van Putten, president of the NWF remarked that the "court has upheld a balanced approach to wolf recovery" and that the "Endangered Species Act has worked to restore a very special part of America's wild heritage and now that success will stand."[3]

The NWF considered the wolf recovery program important not only for its impact on the survival of wolves but also because of its spillover effect. The plight of the wolf in western states is one of many instances of the divisive debate over the Endangered Species Act. Livestock and property rights advocates argue that the act poses economic burdens on them. In contrast, the act is viewed by environmental groups as an important mechanism to protect animal species as well as to help them flourish.

For example, according to one NWF attorney involved in the wolf recovery program, "The 10th Circuit decision means that the Endangered Species Act can be used to restore other species in ways that meet local needs. It can help us unite people to bring back species like the grizzly bear and to stop the decline of others. That's a win for everyone."[4] Moreover, in addition to the efforts promoted by environmental groups, the FWS predicted an increase in park attendance due to the wolf reintroduction plan.[5]

As he neared the end of his term as secretary of the interior, Bruce Babbitt stated that he was "looking forward to visiting Yellowstone one last time to see for myself this program which has been so popular with the public."[6] As to what was accomplished with the program, Babbitt happily stated that "[w]e introduced wolves back into Yellowstone six years ago and it has been one of the most successful actions during my tenure."[7]

During the next dozen years or so, the issue surrounding the place of wolves in the Northern Rockies remained divisive and involved both the federal government and several western states. On the one hand, the FWS under the administrations of both George W. Bush and Barack Obama has pushed a policy of delisting the gray wolf from protections under the Endangered Species Act. On the other hand, Idaho, Wyoming, and Montana have continued to view the gray wolf as a threat to livestock and continue to push efforts to reduce or eliminate recovery of gray wolves. In August 2010, a district court "reinstated protections for all northern Rockies wolves," preventing wolf hunting from going forward in western states, but a short two years later, western states were again pushing wolf management at the state level as they sought to lift protections for the gray wolf, only to have the Center for Biological Diversity file a lawsuit in 2012 to protect the animals.[8]

America's gray wolves are not only part of the western landscape, they are needed to continue to "maintain nature's balance" to ensure that various wildlife herds are "healthy and strong." In short, the various stakeholders—FWS, AFB, environmentalists—must find a way to address the appropriate place of wolves in western states.

Notes

1. See "Restoring Wolves"; http://www.defenders.org/wildlife/ynpfact.html. Accessed January 14, 2001.
2. Ibid.
3. Mark Van Putten quoted in "Court Reversal of Wolf Removal Order: Victory for Common Sense Conservation"; http://www.nwf.org/wolves/court_reversal.html. Accessed March 9, 2000.
4. Ibid.
5. "Restoring Wolves."
6. Office of the Secretary, U.S. Department of the Interior, "Media Advisory: Babbitt Will Visit Yellowstone to Discuss Successful Wolf Re-Introduction"; http://www.doi.gov/news/010105s.html. Accessed January 14, 2001.
7. Ibid.
8. "Saving the Northern Rocky Mountain Gray Wolf"; http://www.biologicaldiversity.org/species/mammals/Northern_Rocky_Mountains-gray-wolf. Accessed May 14, 2004.

"Greenwashing" and Organized Opposition to Environmental Activism

Business and industry have ample resources to use to promote favorable policies that impact their interests. The business sector has employed traditional political methods, including lobbying Congress, working closely with executive agencies, and participating in the litigation process as well as using the airwaves to broadcast commercials that "inform" viewers that business is not only concerned about the environment but it is also doing something about it. For example, during the 2000 presidential primary campaign, the group Republicans for Clean Air, spent more than $2 million on television ads attacking a Republican presidential candidate, Arizona senator John McCain. The donors of the $2 million, as it was later made known, were two Texas brothers, Charles and Sam Wyly, who were supporters of candidate George W. Bush, and whose views on clean air were aligned with Mr. Bush's. The point is, the name of the group was a ruse that misrepresented to the American public its purpose as well as its membership.

More importantly, business and industry have been engaged in numerous examples of "greenwashing," a public relations ploy, according to environmentalists, whereby companies advertise their purported "green" business practices. An example from two decades ago was the National Wetlands Coalition, which was not concerned with protecting the nation's wetlands, but rather was committed to reducing the burden of wetland regulations on real estate and oil and gas interests.[32] Another example is provided by a group of utility companies under the organizational banner of the Endangered Species Reform Coalition, who sought to weaken, not strengthen, the Endangered Species Act. One further example comes from an effort undertaken to protect energy interests of the petroleum industry, united under the Marine Preservation Association, whose name suggested, falsely, a commitment to the marine and ocean environment. More recent examples have included a Christmas tree growing industry that claimed to reduce carbon emissions from the air because of the trees. What this group failed to acknowledge, however, were the dangerous pesticides used to maintain the trees and the carbon emissions generated by harvesting them with helicopters.[33] In 2012, a water bottle company found itself in court because of its questionable use of the word *biodegradable*; according to the Attorney General of the State of California, the bottles failed to biodegrade as required by guidelines set forth by the Federal Trade Commission.

Why might business and industry engage in this type of misinformation? During the 1960s and 1970s, landmark legislation was passed by the Congress and signed by the president in support of environmental protection. Against this background, over the decades, business and industry have felt that their interests were being sacrificed to the interests of environmental protection. According to legal representatives of the business community, the economic sector was caught "in an environmental vise—squeezed by ever-closer scrutiny and harsher penalties while the complexities and breadth of green laws and regulations made full compliance impossible."[34]

Conclusion

In today's American democratic society, citizens communicate their attitudes and opinions through public opinion polls, and attempt to influence public policy through organized interests. Over the last eight decades, surveys have been conducted to ascertain what the public thinks about politics, public issues, political celebrities, and government. Where poll data enable citizens, public officials, academics, business and industry, and organized interests to learn about Americans' political orientation, interest groups have provided an important vehicle to employ as citizens make demands on government.

This chapter has examined public opinion and interest groups and their relationship to the environment as an important public policy issue in American politics. We assessed the public's political attitudes toward a variety of factors relevant to the study of environmental policy. Public opinion polls have informed us that the environment is and remains an important public policy area that should not be neglected by the government or business and industry. American citizens have indicated to policymakers through opinion polls that safeguards on air and water quality, among other environmental protections, need to be maintained and expanded. Policymakers have been encouraged to make a stronger political commitment to environmental protection, by authorizing more funds to be allocated to the environment. Moreover, Americans have exhibited a consistent pro-environment position by affirming in one national poll after another that environmental protection is more important to them than economic growth. Although the public's opinions about environmental affairs have fluctuated over the years, the American people have maintained, overall, strong support for environmental protection.

At the same time, public preferences have been aggregated as inputs into the political process through organized interests. As a characteristic of interest group activity, we find that interest groups exhibit similarities and differences as they employ their resources in support of public policy preferences. During the last century, environmental groups have increased their number, size, and influence upon issues. Environmental activism has become institutionalized in American politics. As we have pointed out, the environmental movement is not united, but is diverse in its goals and methods of operation. While some groups are large and multifaceted, others are smaller and more narrowly focused. The environmental movement has had numerous successes, represented by landmark legislation, presidential actions, and judicial rulings. At the same time, environmentalists will continue to be challenged by recalcitrant public officials as well as by resourceful business and industry actors whose values and economic interests lead them to perceive threats in environmental laws and regulations.

Websites

Public Opinion

Gallup Poll: www.gallup.com

Lewis Harris Center: www.irss.unc.edu/data_archives/pollsearch.html

National Opinion Research Center: www.norc.uchicago.edu

Pew Research Center: www.pewresearch.org

Roper Poll: www.ropercenter.uconn.edu

Yale Project on Climate Change Communication: http://environment.
yale.edu/climate-communication

Interest Groups: Environmental Groups

Defenders of Wildlife: www.defenders.org

Environmental Defense Fund: www.edf.org

National Wildlife Federation: www.nwf.org

Natural Resources Defense Council: www.nrdc.org

Sierra Club: www.sierraclub.org

Wilderness Society: www.wilderness.org

Interest Groups: Business, Industry, Property Rights

American Petroleum Institute: www.api.org

CATO Institute: www.cato.org

Defenders of Property Rights: www.defendersproprights.org

Environmental Working Group: www.ewg.org

Heritage Foundation: www.heritage.org

National Association of Manufacturers: www.nam.org

4

Congress, the Legislative Process, and the Environment

One of the distinctive features of constitutional government in the United States is the separation of powers between the legislative, executive, and judicial branches of government. This division of authority, with each branch checking and balancing the actions of the others, provides for a creative blending and sharing of responsibility for national policymaking. Dividing authority in this way also promotes competition among the branches, each seeking ascendancy, thereby ensuring institutional rivalry, conflict, and inefficiency.[1] The framers deliberately encouraged this interbranch competition because they were anxious to avoid concentrating too much power in any one branch. They also wished to create a government that would be responsive to the diverse interests within the nation.

Congress has an especially important role in making, amending, and rescinding public policy. The policymaking process has been identified as having five stages: problem identification, policy formulation, policy adoption, policy implementation, and policy evaluation.[2] While Congress has a significant part to play in each of these stages, its role is especially important in the policy formulation and adoption stages. It is here that competing proposals for solving public problems are initiated and debated, and coalition building, bargaining, and compromising occur in an effort to build majority consensus around particular initiatives. Congressional appropriations affect agency funding and staffing levels; oversight authority can influence the pace of policy implementation; and hearings and investigations can gather or dispense information to shape agency decisions. Congressional policymaking is affected by structural, operational, procedural, and behavioral factors. To understand congressional decision making requires some familiarity with those factors.

The general structure of Congress is familiar to most Americans, especially its bicameral structure composed of a lower house (House of Representatives), with representation based on population, and an upper chamber (U.S. Senate) whose membership is subject to the principle of equality among states. The larger House, with its 435 members and two-year terms of service, was originally thought to be the more popular and responsive of the two bodies. The one hundred–member Senate, with its six-year terms, was believed to better represent the broader, combined interests of the states and the nation and to be a more deliberative body. While the purported advantages of each body might be disputed, it is clear that the operating environments of the two chambers differ considerably. Some have argued, for example, that the larger size of the House, and the power lodged in certain positions (Speaker of the House) and committees (Rules) impedes its effectiveness relative to the Senate in addressing current problems.

The specific institutional characteristics of Congress are less well known, particularly the committee and subcommittee structure, procedural rules, leadership structure, informal behavior patterns, and constitutional authority. These factors help define the institutional context within which policymaking occurs, and there is considerable variety between the House and the Senate on many of these matters.

Certain characteristics also distinguish environmental policies from those in other policy domains. David Davis identifies five features that set environmental policy apart and make it interesting: high and sustained public interest, issues that elicit emotions or passions, the breadth and scope of environmental effects, the risk to health from inaction or ineffective policies, and the need to understand and manage sophisticated technology.[3]

This chapter examines the role of Congress in environmental policymaking, and some of the factors affecting it. It considers the institutional characteristics of Congress, the congressional arenas most relevant to environmental policymaking, and the formal and informal decision-making mechanisms, processes, and behaviors that shape environmental policymaking. Specific examples are provided to illustrate how these forces influence decisions on environmental matters. Attention is devoted to more recent congressional deliberation on matters of environmental relevance. In recent years, congressional environmental deliberations have resulted in either inaction or minor incremental changes in existing policy rather than the more comprehensive changes sought by environmental advocates. The gridlock and stalemate that currently stymies efforts to pass more encompassing environmental policies of the type approved during the environmental heyday of the 1970s is likely to continue to characterize environmental policy in the near term.

How Congress Influences Policy

Structural Factors

Beyond the general structural factors discussed above that distinguish the U.S. Senate from the House of Representatives, there are more specific structural features that characterize congressional decision making and shape environmental policymaking. Five such characteristics deserve brief attention: fragmentation/dispersion of power, decentralization, multiple checkpoints or avenues of access, parochialism, and short election cycles. Four consequences of these characteristics are important—institutional conflict, incrementalism, gridlock, and the need for integration of policy. Each of these features and its consequences will be considered in turn.

FRAGMENTATION/DISPERSION OF POWER

The environmental policy process in Congress is highly fragmented, with power dispersed widely in several different committees and subcommittees. Nearly two-thirds of the standing committees in the House claim some responsibility for environmental matters. Similar jurisdictional overlap occurs among Senate committees and subcommittees, often leading to rivalry and competition.[4] Jurisdictional division of labor is sometimes based on subject matter (e.g., Senate Committee on Energy and Natural Resources, House Energy and Commerce Committee) and sometimes on broader concerns that include the environment along with other issues (e.g., Senate Homeland Security and Governmental Affairs, House Governmental Operations committees).

The chairpersons of key legislative committees (e.g., Senate Committee on the Environmental and Public Works) and subcommittees (Superfund, Toxics and Environmental Health; Green Jobs and the New Economy; Clean Air and Nuclear Safety; and Oversight and Investigations) are highly influential in advancing or modifying legislation and targeting resources to problems. Woodrow Wilson noted this extraordinary influence a century ago when he described our form of government as "a government by the chairmen of the Standing Committees of Congress."[5] Power in Congress became more fragmented as a result of reforms adopted in 1970 that reduced the power of committee chairpersons and increased that of subcommittee chairs. Nonetheless, today both committee and subcommittee chairs are able to function as environmental policy entrepreneurs, seeking ways to bring together a majority on behalf of (or sometimes against) particular environmental issues and proposals.

DECENTRALIZATION

Because of fragmentation and dispersion of power in Congress, decision making tends to be decentralized. Environmental bills are first crafted and debated in subcommittees, then, if approved, advance to the full committee. Finally, some bills emerge on the floor of the House and Senate for deliberation and vote. Most proposals fail to reach the floor in either chamber. Some have criticized this decentralized decision-making process; others note its advantages.[6] On the positive side, the congressional committee structure provides avenues of access for partisans of various stripes which, coupled with access to the courts and state and local government decision-making bodies, offers numerous venues for various advocacy groups to advance their particular environmental agendas. On the downside, transaction costs are high for legislative leaders who are trying to navigate legislative proposals through the labyrinth of committees and subcommittees. They must avoid becoming ensnared in the complex procedures, running afoul of established norms, and getting entangled in the myriad rules of legislative decision making. House and Senate leaders are not without power in this decentralized environment. They have agenda-setting powers, and their preferences can influence chamber priorities. Their personal skills are also important: to be successful, legislative leaders must be adept at bargaining, compromise, and coalition building.

MULTIPLE CHECKS/ACCESS POINTS

Decentralized decision making and fragmented power offer both advocates and opponents of environmental policy initiatives opportunities to influence policy. Proponents can try to pick the arena in which their proposals are considered, seeking the most congenial environment to enhance prospects for passage of their initiatives. For example, an advocate for change in nuclear energy policy must decide whether to try to have the bill sent for deliberation to the Senate Environmental and Public Works Committee versus the Energy and Natural Resources Committee. Opponents can fashion strategies to kill ill-advised bills and at various stages they are afforded numerous opportunities to defeat them.

PAROCHIALISM

Electoral imperatives requiring legislators to provide particularized benefits to their constituents mean that local and regional concerns often trump national policy considerations. Most environmental policies are of two types: regulatory or distributive.[7] Regulatory policies restrict or extend the options that citizens or business might pursue to achieve their objectives, often through use of sanctions or incentives. Distributive policies confer "particularized benefits" or subsidies upon individuals, groups, or institutions. As Michael Kraft points out, environmental protection policies are usually regulatory, while natural resource and conservation policies are typically distributive.[8] Members of Congress stand to gain political points with their constituents from distributive policies that offer tangible benefits to "the folks back home." Such pork barrel projects provide a powerful motivation for vote-hungry legislators. In the quest for such parochial benefits, legislators often resort to vote trading, or logrolling, offering support for a fellow legislator's priority project in return for the promise of reciprocal support on a project benefiting one's own constituents. For example, in the energy and water area, Senator A might offer support to Senator B for a water project in B's district in return for B's support of a riverfront park in A's district. Policy debates of this type reinforce former speaker of the house Tip O'Neil's well-known observation that "all politics is local."

SHORT ELECTION CYCLES

Members of Congress must keep their eyes on the electoral calendar, particularly those in the House of Representatives, who face voters every two years, and the one-third of the Senate that is elected every two years. Elected officials are forced by condensed election cycles to consider policy proposals from a short-term time perspective. Legislators must be in tune with public opinion; they favor proposals that bring substantial or quick benefits and those with low or deferred costs. Policy proposals are assessed in part based on whether they threaten or enhance the electoral fortunes of legislators. This leads to what Bruntland calls "the tyranny of the immediate," where pressing problems with immediate consequences are given priority over longer-term concerns.[9] The compressed time perspective and preoccupation with reelection, not surprisingly, might lead legislators to defer action on longer-range problems such as global warming and act more quickly on localized problems such as hazardous waste.[10]

INCREMENTALISM

The structural fragmentation within Congress, together with that result-ing from federalism, separation of powers, and bureaucratic fiefdoms, makes integrated policy difficult and incrementalism all but inevitable. Incrementalism is policymaking by small steps—policy change often results in minor modifications or adjustments in the status quo, and less frequently in rapid, comprehensive, or radical change.[11] Congressional policymaking, in the environmental policy area and others as well, often involves carving up policies into manageable pieces and attacking them piece by piece with little attention to the interrelationships among the parts. In the past there has been reliance on command-and-control regulation and technology-specific standards, as well as programmatic emphasis on various environmental areas (e.g., air and water). More recently, other approaches (e.g., market incen-tives, pollution prevention) have been adopted which previously had been infrequent, episodic, and primarily at the margins of public policy.[12]

GRIDLOCK

Policy gridlock is often the result of divided government, partisan bicker-ing, constitutional divisions of power, and interest group maneuvering.[13] Gridlock occurs when different institutions are unable to resolve conflicts, with each blocking the other, preventing the development and enforce-ment of policy. When Congress is unable to resolve conflicts, its inability to act means public problems are not addressed. Divided government and environmental policy gridlock prevailed in the 1980s, 1990s, and the first decade and a half in the 2000s. As a consequence, consensus has been difficult or impossible to achieve on major new environmental initiatives, renewal of existing programs has been problematic, and where policy change has occurred it has typically involved minor modification of past policy. Presidential proposals have been ignored or defeated in Congress, and the president has vetoed congressionally approved initiatives or acted unilater-ally by exercising executive powers. Kraft attributes this policy stalemate to five factors in addition to divided authority and conflicting institutional and political incentives; intractable environmental problems, lack of pub-lic consensus on environmental policy, powerful special interests, high-cost solutions, and absence of effective political leadership.[14]

INSTITUTIONAL CONFLICT

Framers of the Constitution did not seek to avoid institutional conflict in the interest of promoting efficient decision making; they sought instead to ensure conflict and some inefficiency by establishing federalism and the separation of powers in the three branches of government. Institutional conflict was also built into the constitutional relationship between the two chambers of Congress and among the committees and subcommittees in each chamber. It is not surprising that legislative turf battles are commonplace. Fragmented authority has evolved among diverse agencies in the executive branch as well; adding to the conflict is that, once passed, policies must be implemented. Further, competition between the political parties and between the White House and Congress, particularly in the recent years of fierce partisanship, exacerbates conflict and adds to the obstacles facing those hoping to fashion rational comprehensive environmental policies. Frequently, conflict includes disagreement about both the ends (purposes) and the means (strategies) of environmental policy, but in recent years further conflict has resulted because Congress has had the edge over the president in initiating new policies. As noted by Sussman and Kelso (1999):

> Congress, which has overshadowed the president in environmental initiatives, has sought to influence environmental policy via its control over the legislative process and oversight function, especially the authority vested in the powerful committee system.[15]

This congressional "edge" is especially evident when congressional partisanship is strong. However, recently there have been proposals that favor transferring more decision-making power from Congress to the executive branch.[16]

NEED FOR INTEGRATED POLICY

Institutional fragmentation in Congress and other governmental institutions makes it difficult to forge consensus on what to do and how to do it, complicating the task of achieving integrated environmental policy.[17] The overlapping legislative committee jurisdictions and semiautonomous subcommittee activities involve numerous policymakers, each with their own agenda, resources, and prerogatives and operating with little or no coordinating effort. Congressional deliberations are also influenced by iron-triangle or issue-network relationships with administrative agencies and environmental interest groups (see chapter 6). These subgovernments angle for approval or defeat of legislative policies that promote or do not impede their organizational interests.[18] Such networks improve prospects for communication, bargaining, and consensus building among diverse interests; however, each participant has a vested interest and jurisdictional turf to protect and may resist proposals for change that threaten their interest, turf, or primacy in the policy process.

Operational Characteristics

While these structural features and their consequences are important in understanding the institutional context of congressional decision making, certain operating mechanisms are also significant—specifically, the role of authorizing statutes, appropriations processes, and oversight activities. The success or failure of policy formulation and implementation often depends on legislative actions in these areas.

AUTHORIZING STATUTES

Enabling legislation not only sets broad goals and standards for environmental policies, it also specifies how those goals are to be implemented, sometimes in considerable detail. For example, such laws might specify that risk-based approaches be used or that technology-based strategies be undertaken. The extent to which costs must figure into the decision differs from statute to statute. National policy goals are sometimes articulated in enabling legislation; other times, goal setting might be reserved for state action. Daniel J. Fiorino discusses this variation in the content of authorizing statutes indicating that some laws specify the medium to be used in reducing pollution, others the sector (public versus private) responsible for funding, and still others the impact of pollution (e.g., acid rain).[19] Congress clearly sets the direction of environmental policy by crafting authorizing legislation.

APPROPRIATIONS

An alternative stream in the policy process involves the allocation of resources to carry out environmental policies. An important distinction is the role of congressional authorization and appropriation committees. Authorizing committees give agencies the legal authority to operate and specify the funding levels; appropriations committees give agencies the authority to spend money on environmental problems. Authorizing committees (i.e., standing committees) have close ties to the agencies they oversee. Influential appropriations committee and subcommittee chairs and ranking members can channel funds to projects and locales that coincide with their priorities. Riders to appropriation bills can fund studies or earmark funds to pet projects. Environmental interest group leaders and their counterparts from business and industry know that it is important to stay on the good side of influential appropriations committee members; astute administrative agency leaders are aware of this requirement as well, especially if they want agency monies targeted to specific problems.[20]

OVERSIGHT

We have mentioned the important congressional committees that engage in oversight in both the House and Senate. In chapter 2, oversight was also considered as one of the regulatory "sticks" that the federal government uses in its relations with states and localities. While oversight can and sometimes does lead to extensive control (critics refer to this as micromanagement), it can also be loose and ad hoc, depending on the preferences of those on the congressional committees and subcommittees. Oversight can facilitate or impede administrative change. Committee chairs often determine the scope and intensity of oversight activity. They can also influence which departments and agencies are responsible for particular programs. In some instances, a broad range of oversight tools might be used—requests for agency testimony, congressional letters of inquiry, investigations, audits, reports, political pressure on agency personnel, and the like.[21] Agencies often object to intrusive congressional oversight, viewing it as disruptive and threatening to their autonomy. In response, they may attempt to manage Congress by developing close ties with the committee and subcommittee chairs and staff and cultivating a favorable view of agency purposes and programs. Sometimes agency heads and committee or subcommittee chairs jointly criticize agency performance.

Processes and Behavioral Characteristics

The structural characteristics and operational mechanisms in the legislative branch create and shape the environment in which public policy is made and carried out. A more specific examination of the policy process, especially as it influences environmental policy, involves examination of the modes of decision making and behavioral dynamics found in the legislative arena. Here the focus is on the nature of decision processes (detailed policy guidance, micromanagement) and links to relevant constituencies (partisanship, group pressures). Also considered is the political maneuvering (credit claiming), symbolic and self-serving politics (symbolism/policy layering, pork barreling, nongermane riders), and additional features tied to environmental policymaking (reactive decision making, science and politics, competing values, and public interest). These processes and behavioral characteristics, together with the structural features and operating mechanisms previously considered, provide additional insight into the legislative process and the environmental policies that emerge.

DETAILED POLICY GUIDANCE

As noted in chapter 6, administrative agencies derive considerable power when they are given substantial discretionary authority. In recent years, Congress has reduced administrative discretion by specifying in considerable detail the standards, deadlines, and regulatory mandates. This leaves little "wiggle room" for administrators, who are expected to toe the line in conformance with congressional directives. In part, this move away from delegation to more controlling policy guidance is a reflection of the growing distrust between Congress and the White House regarding environmental policy—another consequence of divided government. The use of inflexible language and detailed prescriptive regulations increases the costs of environmental regulation. It does, however, preserve congressional prerogatives to establish environmental policy, reduce the likelihood of poor administration, and enable aggrieved parties to bring legal action when agency administrators do not meet their responsibilities.[22] The flip side sometimes occurs as well: Congress may approve statutory language that is ambiguous, vague, and subject to multiple interpretations, resulting in part from legislators' inability to achieve consensus. In such instances it is left to the administrative agencies and courts to resolve the ambiguities through bargaining or litigation.[23]

MICROMANAGEMENT

In an effort to curb administrative discretion, particularly at the Environmental Protection Agency, Congress has tightened its legislative controls.[24] Ironically, such controls have been encouraged by both pro-industry congresspersons and by pro-environmental legislators—each seeking to influence EPA implementation and enforcement activities—one side seeking implementation delays and enforcement that is relaxed or sympathetic to business concerns, and the other side seeking to carry out legislatively mandated regulations in a timely and aggressive manner. These conflicting perspectives have led to battles over legislative language and resource allocation. As legislative guidance becomes more specific and controls tighten, administrative flexibility and discretion decrease, sometimes resulting in implementation failures or delays, reduced agency credibility, and increased litigation.[25] Those who view this trend favorably argue that tightened congressional control strengthens the hand of agency administrators by providing a counterweight to other influential agencies such as the Office of Management and Budget; those with less favorable views question the extent of helpful guidance or support received by agencies from such close oversight.[26]

Personal Profile

Senator Susan Collins: Environmental Defender

"Global climate change is the most significant environmental challenge facing our nation today, and we must develop reasonable solutions to reduce our greenhouse gas emissions."[1] This quote comes from a letter written to constituents by an elected public official. It was not penned by a liberal Democrat in the U.S. Senate, but by four-term senator Susan Collins, Republican from Maine. She was elected to the Senate in 1996, reelected in 2002 (with 59 percent of the vote) and 2008 (with 61 percent of the vote), and recently again in 2014 (with 68.5 percent of the vote).[2]

Collins is a magna cum laude graduate of St. Lawrence University, where she was elected to membership in Phi Beta Kappa, the national academic honor society. She spent a dozen years working on the Capitol Hill staff of Senator William Cohen, followed by another five years as a cabinet member of Governor John McKernan, where she served as commissioner of professional and financial regulation. Her experience also includes two years in the early 1990s as New England Administrator of the U.S. Small Business Administration. She made history in Maine as the first woman to receive a major party nomination for governor in her ultimately unsuccessful bid for elective office in 1994. Shortly after her defeat, she became the founding executive director of the Center for Family Business at Husson University in Bangor, Maine. Her successful candidacy for the U.S. Senate followed in 1996.[3]

Currently, she is the ranking member and former chairperson of the Homeland Security and Governmental Affairs Committee. She supports a national strategy for achieving energy independence by 2020, government efforts to create green energy jobs, and reduction in U.S. dependence on foreign oil. She co-sponsored and introduced bipartisan legislation, the Carbon Limits and Energy for American Renewal (CLEAR) Act, a clean energy and climate bill containing a cap-and-dividend that would promote energy technology and financially benefit consumers. Collins is a strong advocate for land conservation, efforts to protect air quality, and deep-water offshore wind energy research and development, among other things. She was able to secure a $20 million commitment from the secretary of energy to invest in clean energy technology.[4]

When congressional oversight hearings on the BP Gulf of Mexico oil spill were held, they were conducted under the auspices of the Senate Homeland Security and Government Affairs Committee, where she is ranking member. She has forcefully favored reforms to ensure that adequate preparations are made to avoid future disasters. Working with Senator Joseph Lieberman (ID-CT) she co-authored legislation to reorganize the Federal Emergency Management Agency (FEMA) within the Department of Homeland Security, which, among other things, strengthens FEMA's preparedness and response capabilities. Nonetheless, she broke ranks with environmentalists on one important recent issue, the Keystone XL pipeline, where she argued, "[W]e need that oil."[5]

Her environmental leadership overall has been recognized by the League of Conservation Voters (LCV) who not only endorsed her reelection bid—the only Republican that LCV endorsed in the 2008 Senate race—but also recognized her with a perfect score on their National Environmental Scorecard the year before.[6] LCV's senior vice president Tony Massaro has been quoted as saying, "Senator Collins is a leader and champion for the environment, reaching across party lines to introduce and support bipartisan pieces of legislation that hold oil companies accountable for their high profits, invest in clean energy alternatives, and promote fuel efficiency for vehicles."[7] Massaro also praised her leadership and her "amazing ability" to make the case for strong global warming legislation both within her own party and across the aisle to wavering Democrats. David Jenkins, the government affairs director of Republicans for Environmental Protection, praised her as an "outspoken champion of taking on climate change."[8] His group named Senator Collins "Greenest Republican" for the second time in 2010, while simultaneously chastising less environmentally friendly "Brown Republicans."[9]

Senator Collins's straight-talking, bipartisan, moderate approach has enabled her to establish a solid track record of environmental leadership. The scope of her public policy involvement extends well beyond her work on behalf of the environment. For example, in addition to her service on the Homeland Security and Governmental Affairs Committee, she sits on the powerful Appropriations Committee, Armed Services Committee, Special Committee on Aging, and her prior service includes membership on the Committee on Health, Labor and Pensions as well as leadership of the Permanent Subcommittee on

Investigations.[10] Many professional and civic groups have honored her, bestowing such titles as "Ports Person of the Year," "Guardian of Small Business," "Legislator of the Year," "Public Service Award" recipient, and highlighting her distinguished service.[11] The work of Senator Collins demonstrates that even in a hyperpolarized political arena, committed individuals regardless of political party affiliation are working tirelessly to advance the public interest and protect the environment.

Notes

1. Grist staff, "Tracking Where Senators Stand on Climate Legislation: Susan Collins (R-Maine) [Updated]," *Grist*, December 10, 2009; http://grist.org/article/2009-susan-collins-on-climate-legislation/. Accessed September 26, 2014.

2. Biography staff, "Sen. Susan Collins," *Alliance To Save Energy*; https://www.ase.org/biography/susan-collins. Accessed September 26, 2014.

3. Ibid.

4. "Speakers: Senator Collins," Energy is Urgent, Washington Post Live forum event, *The Washington Post Live*, September 23, 2010; http://washington postlive.com/conferences/speakers/senator-susan-collins. Accessed September 26, 2014; PR staff, "Senator Collins 'Greenest Republican' in Senate," *Susan Collins United States Senator for Maine*, July 24, 2010; http://www.collins.senate.gov/public/index.cfm/press-releases?ID=05b54b06-802a-23ad-4917-4231f8d334b7. Accessed September 26, 2014.

5. Linda Dumey, "Letter to the Editor: Sen. Collins Favors Tar Sands over the Environment," *Portland Press Herald*, March 5, 2014; http://www.pressherald.com/2014/03/05/letter_to_the_editor__sen__collins_favors_tar_sands_over_the_environment/. Accessed September 26, 2014.

6. Kate Sheppard, "LCV Endorses Republican Susan Collins in Maine Senate Race," *Grist*, October 17, 2008; http://grist.org/article/maine-dish/. Accessed September 26, 2014; LCV staff, "2013 National Environmental Scorecard: Senator Susan Collins (R)," *League of Conservation Voters*; http://scorecard.lcv.org/moc/susan-m-collins. Accessed September 26, 2014.

7. Kate Sheppard, "LCV Endorses Republican Susan Collins in Maine Senate Race," *Grist*, October 17, 2008; http://grist.org/article/maine-dish/. Accessed September 26, 2014.

8. Dennis Sanders, "The Greenest Republicans?" The Moderate Voice, April 23, 2009; http://themoderatevoice.com/29765/the-greenest-republicans/. Accessed September 26, 2014.

9. Ibid.

10. Biography staff, "Sen. Susan Collins."

11. "Senator Susan Collins of Maine Recognized by National Park Trust," *National Park Trust*; http://www.parktrust.org/about-us/awards-and-events/bruce-f-vento-award/71-national-park-trust/about/bruce-f-vento-past-recipients/333-senator-susan-collins-of-maine-recognized-by-national-park-trust. Accessed September 26, 2014; Biography staff, "Sen. Susan Collins."

PARTISANSHIP

Environmental policy issues often arouse partisan passions among legislators. While legislators in both parties like to be viewed as pro-environment, the votes on key issues indicate clear differences between Democrats and Republicans. The bipartisan League of Conservation Voters (LCV) and the Environment America (EA) issue national environmental scorecards rating legislators each year. In 2013 Senate Republican chairs had an average, LCV score of 23 and an average EA score of 0, compared to the ranking Democrats' scores of 95 and 100, respectively. For Democrats the chairs' average is above the average for their party (LCV = 92; EA = 91), and for Republicans the chairs' LCV average is above the average for their party (LCV = 17), while the EA average (EA = 5) is below that of the party. The average for Republican chairs is below the average for the Senate as a whole (LCV = 58; EA = 52) on both ratings; conversely, the average score for ranking Democrats is above the average for the Senate on both measures. A similar pattern is evident in the House, where the Republican chairs' average LCV score is 7 and the average EA score is 0, while the ranking Democrats' score LCV score is 72 and the average EA score is 67; in each case these scores are below the average for their respective party. The Democrats' LCV and EA scores, and those of their ranking members, are well above those of the corresponding Senate and House averages, and the Republicans' LCV and EA scores and leadership ratings are considerably below those of the average in each chamber.

Partisanship does make a difference in legislative voting behavior on environmental issues. Clearly, the Democrats have a more pro-environment voting record on these issues than their Republican conterparts.[27] Stylistic differences also exist between Republicans and Democrats on environmental matters. Democrats are more likely to favor activist government, support intervention in business, and favor command-and-control regulation to achieve environmental objectives. Republicans rely more heavily on market-based approaches and seek alternatives to traditional command-and-control strategies. More moderate legislators, such as Congressman Mark Kirk (R-IL) and Senator Susan Collins (R-ME), attempt to balance two seemingly irreconcilable objectives: being a good Republican and a good environmentalist, a tricky balancing act.

Nonetheless, Davis points out that when times change and their electoral prospects require it, parties and politicians can change their colors quickly.[28] President George H. W. Bush and Republicans championed a cap-and-trade approach to cut sulfur dioxide, which causes acid rain, as a business-friendly, market-based approach, then decades later rejected it when President Barack Obama supported it as a way to limit carbon emissions from power plants (Snyder and Martin 2014; Conniff 2009).[29]

Case Study

Cap-and-Trade: Failure to Pass a Climate Bill

The idea that people can buy or sell the right to pollute has been variously referred to as "cap-and-trade" or "emissions trading." This was widely viewed as a way to bring together business-oriented free-market conservatives and moderate to left-leaning environmentalists. Building on early bipartisan legislative successes in the 1990s designed to curb power plant pollution causing acid rain, it was later thought that the same market-based cap-and-trade strategy could be used to reduce carbon dioxide emissions linked to climate change. Instead of using command-and-control mandates to limit pollution, the idea was to impose a cap on emissions, provide polluters with an allowance of pollutants they could emit up to a certain level, give the polluter discretion in how to use its "right to pollute," and allow polluters to sell unneeded allowances or buy additional allowances on the open market, as necessary.[1] While this approach was successful in fixing the acid rain problem, and despite growing bipartisan support between 2003 and 2008, it failed to gain congressional approval in 2009–10 as a strategy for addressing the larger challenge of global warming.[2]

Several explanations have been given for the failure of cap-and-trade to be adopted as national climate policy. Among the most prominent explanations are the following: mistrust of the method by government regulators, skepticism among key environmentalists and congressional subcommittee members, lack of consensus among White House staffers, a faulty game plan by the coalition of business and environmental leaders pushing cap-and-trade, Tea Party protests, lack of citizen understanding and support for the strategy and their concern about increased costs and regulations, the deep recession, complexity of the legislation, fragmentation of the environmental community, substantial funding by those opposing the legislation (oil, coal, and other energy interests), hostility to the science of global warming, demonization of the legislation as "cap and tax," explosion of the British Petroleum oil rig in the Gulf of Mexico, the victory of Republican Scott Brown in the Massachusetts Senate special election, President Obama's unwillingness to push harder for cap-and-trade and priority given to health care reform, and moderate Republicans withdrawing their support.[3] By July 2010, Senate leader Harry Reid decided not to bring the cap-and-trade bill to the Senate floor, concluding it could not pass the sixty vote threshold needed to overcome the filibuster bar.[4]

In a detailed 142-page dissection of the politics surrounding the failure of cap-and-trade legislation, and drawing comparisons with comprehensive health reform legislation, Theda Skocpol summarized and assessed many of the explanations above and concluded that several postmortems are incorrect and others fail to focus on key changes that sealed the fate of the climate legislation.[5] She identifies the pivotal role of the "rightward-lunging Republic Party" fomenting Congressional partisan polarization and grassroots agitation by ultraconservative groups, together with the inability of insider "grand bargaining strategy" of principal proponents of climate legislation to nudge through reform legislation. Partisan gridlock and other factors make the likelihood of further federal action on comprehensive climate change bleak in the immediate future, although the Obama administration is offering incentives for states to act alone in developing their own state and regional carbon-trading systems modeled on those in California and in the Northeast.[6]

Notes

1. Richard Conniff, "The Political History of Cap and Trade," *Smithsonian Magazine*, August 2009; http://www.smithsonianmag.com/air/the-political-history-of-cap-and-trade. Accessed October 30, 2014.

2. Theda Skocpol, "Naming the Problem: What It Will Take to Counter Extremism and Engage Americans in the Fight against Global Warming," paper presented at the symposium on The Politics of America's Fight Against Global Warming, Harvard University, Cambridge, Massachusetts, February 14, 2013.

3. Suzanne Goldenberg, "Climate Change Inaction the Fault of Environmental Groups, Report Says," *The Guardian,* January 14, 2013; http://www.theguardian.com/environment/2013/jan/14/environmental-groups-climate-change-inaction. Accessed October 30, 2014; Miles Mogulescu, "Can We Solve the Climate Crisis If We Don't Solve the Democracy Crisis?" *Huff Post Politics*, February 6, 2013; http://www.huffingtonpost.com/miles-mogulescu/can-we-solve-the-climate-_b_2631033.html. Accessed January 7, 2014; Ryan Lizza "As the World Burns: How the Senate and the White House Missed their Best Chance to Deal with Climate Change," *The*

New Yorker, October 11, 2010; http://www.newyorker.com/magazine/2010/10/11/as-the-world-burns. Accessed October 30, 2014; Eric Pooley, "Why the Climate Bill Failed: It's Not that Simple," *Grist,* January 18, 2013; http://grist.org/climate-energy/why-the-climate-bill-failed-its-not-that-simple/. Accessed October 30, 2014; Brad Plumer, "Why Has Climate Legislation Failed? An Interview with Theda Skocpol," *The Washington Post,* January 16, 2013; http://www.washingtonpost.com/blogs/wonkblog/wp/2013/01/16/why-has-climate-legislation-failed-an-interview-with-theda-skocpol/. Accessed October 30, 2014; Steven Cohen, "The Real Climate Choice Before Congress: Cap and Trade or Command and Control?" *Huff Post Green,* January 5, 2010; http://www.huffingtonpost.com/steven-cohen/the-real-climate-choice-b_b_411662.html. Accessed January 7, 2014; Skocpol, "Naming the Problem"; Lee Wasserman, "Four Ways to Kill a Climate Bill," *New York Times,* July 25, 2010; http://www.nytimes.com/2010/07/26/opinion/26wasserman.html?_r=0. Accessed October 30, 2014; Daniel J. Weiss, "Anatomy of a Senate Climate Bill Death," *Center for American Progress,* October 12, 2010; http://www.americanprogress.org/issues/green/news/2010/10/12/8569/anatomy-of-a-senate-climate-bill-death/. Accessed October 30, 2014.

4. Skocpol, "Naming the Problem."

5. Ibid.

6. Ibid.; Jim Snyder and Martin Christopher, "Politics May Sour Cap-And-Trade Sweeteners in Obama Plan," *Bloomberg,* June 3, 2014; http://www.bloomberg.com/news/2014-06-03/politics-may-sour-cap-and-trade-sweeteners-in-obama-plan.html. Accessed September 29, 2014.

GROUP PRESSURES

Pressures from environmental advocacy groups and industry interests often exacerbate conflict and hamper consensus in Congress.[30] For example, reformers point to the need for more integrated environmental policy and management, but Walter A. Rosenbaum points out that privileged interests represented in environmental subgovernments have undermined or resisted such efforts.[31] Just as organized groups resist change, they can also promote it. Environmental groups are often strategically poised to capitalize on dramatic events or "crisis" situations, to press for policy changes that address the problem in question. For example, massive oil spills and rivers catching fire prompted advocacy groups to call for legislative solutions. Absent such drama, environmental groups will seek other ways to move lower-salience issues (such as the dangers of radon exposure) to greater visibility by effective use of media, lobbying, and public relations. However, some observers have questioned whether environmental groups have the political resources necessary to exert significant influence in the legislative process.[32]

Pro-business groups are equally or more adept at drawing attention to their concerns as they seek support for initiatives such as regulatory reform in an effort to reduce the compliance costs they view as burdensome. With the multiple avenues for access discussed previously, there are ample opportunities for resource-rich groups to articulate their preferences to members of Congress. For example, pro-development forces underwritten by resource-extraction industries, and pro-preservation interests who want protection whatever the costs, compete to have their voices heard and heeded in the legislative arena. Iron triangles and diffuse networks facilitate this communication for those who are included in these "subgovernments" (see chapter 6). Also, groups with money to give legislators to aid in their reelection efforts are well positioned to have their policy preferences considered.

CREDIT CLAIMING AND PORK BARRELING

Members of Congress have a continuing interest in their own reelection prospects. Research on congressional self-interest notes the connection between a congressperson's pursuit of legislative policy goals and his or her overriding desire to win reelection.[33] An important avenue to reelection is showing voters what you have done to make their lives better. "Credit claiming" is a way for a legislator to convince voters that he or she, as an elected representative, is doing something to improve the constituents' life. Such improvements may include pork barreling—providing projects, grants, and contracts that benefit the home constituents. David Mayhew points out that providing specific, tangible, localized benefits and making clear the representative's active role in securing these benefits provides a possible electoral payoff for the legislator.[34] In the environmental protection domain, Michael Lyons notes that there are numerous potential particularized benefits that legislators can distribute, such as federal water projects, sewage waste treatment construction grants, and designation of new parks.[35] Credit claiming can also occur as legislators work to safeguard constituent interests by protecting jobs or preserving health. Elected officials are eager to provide such environmental pork barreling to their constituents and to claim credit for their efforts. As Lyons observes, "The U.S. political system offers to politicians abundant incentive to provide tangible and specific policy benefits, yet relatively little incentive to provide benefits that are diffuse or intangible."[36]

RIDERS

In recent years, efforts to reauthorize major environmental legislation such as the Endangered Species Act and the Superfund law have been unsuccessful. This has led to the question of whether Congress has either the will or capacity to engage in environmental lawmaking of the type approved in earlier decades.[37] The inability to pass major environmental legislation has led to the use of stealth strategies by some. A popular tactic used by anti-environmental legislators has been to attach riders to appropriation bills. A rider is like a "hitchhiker on a freight train," an amendment to a bill that is not germane to the bill's purpose and may be used to restrict, redefine, substantially modify, or cease operations of another, unrelated federal program. Some of these riders are in excess of one hundred pages long. Riders are often a ploy used by Congress to gain leverage over White House reservations about a policy. When Congress attaches riders to funding bills, for example, it becomes necessary for the president, lacking a line-item veto, to veto the entire spending bill or approve it with the unpalatable amendments.[38]

Some observers have predicted that instead of comprehensive reform legislation, the future is likely to see increased use of stealth strategies by those seeking to weaken environmental requirements. Piecemeal change through appropriations riders or other forms of nongermane legislation may be used to circumvent debate and minimize accountability for decisions. However, those preferring more straightforward legislative strategies suggest that this "stealth" trend is unlikely to succeed indefinitely. Jonathan Cannon argues that government agencies and environmental advocates are alert to the dangers of piecemeal approaches and likely to respond aggressively when they are tried, and that fundamental policy issues are too visible, and environmentalists too vigilant, to be continually compromised by stealth strategies.[39]

SYMBOLISM, POLICY LAYERING, AND BLAME AVOIDANCE

One way that members of Congress can show their interest in environmental policy is to engage in policy symbolism. Policy symbolism takes two forms—namely, congressional resolutions that lack legal force but express legislative concern about a problem or issue, and legislative actions that clarify policy goals and have potential impact, but which are intended to fall short of goal accomplishment.[40] Legislators might try to claim credit and avoid responsibility for policy failure by taking symbolic actions. For example, they could specify the goals or aspirations for a policy and then either fail to allocate the necessary resources for it to succeed, or deliberately complicate enforcement or implementation, attributing subsequent problems to bureaucratic inaction.[41]

Lyons alludes to policy layering as another form of symbolism whereby contradictions exist between new and existing policy goals. He notes examples of instances where layering of new policy objectives on top of existing ones renders the accomplishments of initial objectives impossible. Policy symbolism and layering may, in some instances, placate environmental interest groups seeking reassurance and support. Blame avoidance is a strategy used by legislators to mask their actions and put the blame on others (e.g., declining to impose regulations but instead delegating that task to administrative agencies).[42]

REACTIVE DECISION MAKING

Congressional decision making on environmental matters is often in response to a dramatic event or perceived crisis. Reacting to fluctuations in public opinion, legislators often display what Rosenbaum refers to as "'pollutant of the year' mentality." For example, problems such as Love Canal and the chemical explosion in Bhopal, India, provided impetus for passage of Superfund legislation and its amendments; the *Exxon Valdez* oil spill spurred Congress to pass the Oil Pollution Prevention Act of 1990.[43] However, dramatic events don't always result in approved legislation. For example, there was no major legislative response to the BP Gulf of Mexico oil spill (April 20, 2010). Nonetheless, reactive decision making does occur frequently and, as Rosenbaum observes, it "ensures an environmental agenda in which place and priority among programs depend less on scientific logic than on political circumstance. Often the losers are scientifically compelling environmental problems unblessed with political sex appeal."[44] Reactive decision making also leads to policy implementation difficulties because often agencies are given new, untested programmatic responsibilities, one on top of the other, with little advance notice and with insufficient resources to effectively carry out their new tasks.[45]

SCIENCE AND POLITICS

Scientific and technical issues require a level of sophistication that is often lacking among "scientific amateurs" in Congress, many of whom typically come from a legal or business background. This is certainly true on such environmentally complex matters as ozone depletion, nuclear energy, and pesticides. Congresspersons often have neither the time nor the expertise to tackle such difficult and intricate problems, preferring instead to shift such matters to staff, professionals, or interest group experts. However, a distrustful Congress may be reluctant to grant administrative agencies the flexibility they need on scientific matters to effectively provide environmental protection.[46] Also, legislators may be skeptical about scientific claims; for example, many in the GOP dispute the science of climate change.

COMPETING VALUES

As noted above, debates concerning environmental policy often involve discussion of competing values such as environmental protection, economic concerns, and equity considerations. Environmental advocates are often on the opposite side of issues from business and industry interests. Not surprisingly, pro-business concerns often revolve around economic preoccupation with the costs of complying with "burdensome" government regulations and the desire to ensure that benefits exceed costs before such regulations are imposed on business. Environmentalists are much more inclined to identify the benefits derived from environmental protection measures and to understate or discount the importance of the cost side of the equation. Government interests must also examine legislative proposals with an eye on equity considerations, seeking to craft legislation that will take into account fairness issues (e.g., environmental justice policy) together with environmental protection and economic concerns. Because this juggling act of trying to balance tradeoffs among competing values is difficult, policies addressing such values are often inconsistent. For example, some laws may authorize agencies to issue and enforce environmental standards, with no mention of cost considerations. Others mandate action regardless of cost. Still others require considerations of cost without requiring similar attention to benefits, or mandate assessment of both costs and benefits.[47]

PUBLIC INTEREST

Given the structural characteristics, the operating environment, and the process/behavioral patterns cited above, it is a wonder that congresspersons do approve environmental policies that advance broader public interest concerns. James Q. Wilson uses a four-part framework to examine public policy, focusing on the distribution of costs and benefits. His typology includes majoritarian politics (widely distributed costs and benefits), interest group politics (narrowly concentrated costs and benefits), entrepreneurial politics (narrowly concentrated costs, widely distributed benefits), and client politics (widely distributed costs, narrowly concentrated benefits).[48] There are some instances in which a relatively unorganized public wins a policy fight against organized interests. Using Wilson's typology this occurs in entrepreneurial politics where the benefits accrue to a broad public constituency and the costs are imposed on a narrow organized interest. Research by Wilson and by R. Douglas Arnold suggests that prevailing in such battles requires congresspersons who are skilled and dedicated, issues that capture that attention of the public at large, and incentives in the form of electoral payoffs as inducements for the congressional "policy champion" or entrepreneur.[49]

It is clear that members of Congress vary in their motivation to serve the public interest. As Richard Andrews has observed, "Members of Congress may be statesmen seeking the long-term good of the society, decent but parochial representatives of their constituents, or merely self-interested incumbents selling themselves to interest groups to finance their own reelections."[50] Getting the statesmen, the home-style legislators, and the self-promoting careerists to agree on environmental policy is a major challenge and requires discovery of common ground that simultaneously serves a combination of public and private interest objectives.

Conclusion

The flurry of legislative activity during the environmental decade of the 1970s created the foundation for the more incremental modifications in policy that have followed in subsequent decades. It is unlikely that in today's political environment there will be comprehensive environmental initiatives put forward by Congress and approved by the president. With the election of Barack Obama and near-parity between the parties in Congress, it is more likely that environmental policy in the near term will be incremental rather than reflect new policy initiatives.

What is the prognosis for future legislative action on environmental policy? The preceding analysis of structural factors, operational characteristics, and behavioral/process features provides some clues. The structural factors are intractable and the consequences are predictable. The institutions of government were designed to disperse power; this fragmentation is evident in Congress, with its two chambers and decentralized decision centers providing access to numerous competing interests. Short election cycles and the parochial focus on reelection will continue to influence legislative deliberations regardless of future partisan electoral outcomes. These structural factors guarantee institutional conflict and tilt the system toward incrementalism and short-term thinking, and away from integrated approaches to policy problems. Electoral outcomes may well determine the extent to which policy gridlock continues to accurately describe congressional decision making. Like structural characteristics, operating mechanisms are unlikely to change in substantial ways. Congress will continue to make policy by authorizing statutes, appropriating funds, and overseeing agency implementation. The number and type of statutes approved, the size of appropriations, and the extent of oversight might change based on future election outcomes, but the mechanisms will still operate in much the same way before and after the votes are counted.

It is in the behavioral/process area where changes are most likely to occur. Detailed policy guidance and micromanagement result in part from breakdown in trust and from policy differences between various participants in the policy process, most notably the White House and executive agencies on the one hand and Congress on the other. Were the same political party to control both the White House and the two chambers of Congress, the trust gap and policy differences would likely shrink, along with many of the incentives for detailed policy guidance and micromanagement. Competing perspectives on environmental policy will continue to divide Democrats and Republicans as well as environmental advocates and business/industry partisans. Those favoring a legal/regulatory approach emphasizing absolute standards and de-emphasizing compliance costs will continue to be pitted against those preferring an economic strategy emphasizing market mechanisms and de-emphasizing political interference. These cleavages make environmental policy conflict inevitable, but do not preclude the possibility of bipartisan cooperation. The reelection imperatives of members of Congress make credit-claiming riders and pork barreling indispensable, and there is no reason to believe that symbolism, policy layering, and blame avoidance will take on less importance in the future. Similarly, legislators are likely to continue to engage in reactive decision making, but opportunities do exist for more proactive policy entrepreneurs to strike out in new directions to advance pro-environment agendas.

Future congressional deliberations on environmental matters will undoubtedly involve questions about the quality of scientific evidence supporting specific policy proposals, the extent of the risk for action or inaction, and the likelihood that the proposed solutions will actually ameliorate the targeted problems. Those pushing an environmental agenda will need to marshal facts to support claims about environmental "threats" and to make the case that limited resources should be spent on environmental protection rather than on other, competing problems. Value conflicts will persist as legislators seek to maintain a balance among a trio of environmental, economic, and equity concerns. Congressional deliberations will continue to reflect the mixed motives of legislators themselves; public interest, self-interest, and parochial representation interests will simultaneously be weighed and traded off against each other. Amid these competing scientific claims, conflicting values, and mixed motives, there is some consensus that environmental policies require greater flexibility and that much more attention be given to weighing the costs and benefits of policy alternatives.

Environmental protection and cost-effective policies are not mutually exclusive. Wasteful spending can be avoided without rolling back environmental protection. While environmental policy debates in the past have been "confrontational in style and polarizing in practice,"[51] this does not have to be the case in the future. Politically savvy and skilled legislative leadership, an aroused and active public, and an attractive set of policy proposals with appeal to the electorate are needed to build on this consensus and to take constructive action. Such conditions are necessary if comprehensive new environmental initiatives are going to successfully weave their way through the labyrinthine legislative obstacle course.

Websites

Policymaking Process Stages

http://www.ushistory.org/gov/11.asp

Congress Policymaking

https://www.congress.gov/legislative-process

The Legislative Process

http://www.house.gov/content/learn/legislative_process/

House

http://www.house.gov/

Senate

http://www.senate.gov/

Latest Environmental Policy News

http://www.sciencedaily.com/news/earth_climate/environmental_policy/

http://www.nrdc.org/legislation/

Environmental Committees and Subcommittees' Webpages Directory

http://www.congressmerge.com/onlinedb/cgi-bin/committee_list.cgi?site=
congressmerge

Specific Environmental Committees and Subcommittees

U.S. SENATE COMMITTEE ON ENVIRONMENT AND PUBLIC WORKS

http://www.epw.senate.gov/public/

SUPERFUND

http://www.epa.gov/superfund/

AGENCY FOR TOXIC SUBSTANCES AND DISEASE REGISTRY

http://www.atsdr.cdc.gov/

NATIONAL CENTER FOR ENVIRONMENTAL HEALTH

http://www.cdc.gov/nceh/

U.S. SENATE COMMITTEE ON ENVIRONMENT AND PUBLIC WORKS

http://www.epw.senate.gov/public/

OVERSIGHT AND INVESTIGATIONS

http://energycommerce.house.gov/subcommittees/oversight-and-investigations

Overview of the Authorization-Appropriations Process

http://www.senate.gov/CRSReports/crs-publish.cfm?pid='0DP%2
BPLW%3C%22%40%20%20%0A

Senator Susan Collins

http://www.collins.senate.gov/public/

Democrats vs. Republicans in Environmental Policy Issues

http://www.usnews.com/news/blogs/data-mine/2015/07/06/democrats-
republicans-disagree-on-energy-global-warming

http://www.pewresearch.org/daily-number/many-more-democrats-than-
republicans-say-protecting-environment-a-top-priority/

http://classroom.synonym.com/green-energy-views-democrats-vs-
republicans-5534.html

http://www.gallup.com/poll/107569/climatechange-views-republican
democratic-gaps-expand.aspx

League of Conservation Voters

http://www.lcv.org/

Environment America

http://www.environmentamerica.org/

Environmental Pressure Groups

http://www.sierraclub.org/
http://www.alleghenylandtrust.org/
http://climaterealityproject.org/
http://www.citizenscampaign.org/

http://www.earthfirst.org/

http://www.environmentamerica.org/

http://www.earthshare.org/

https://www.treepeople.org/

http://www.conservationfund.org/

Exxon Valdez Oil Spill

http://www.evostc.state.ak.us/index.cfm?FA=facts.QA

BP Gulf of Mexico

http://ocean.si.edu/gulf-oil-spill

5

The Environmental Presidency

Introduction

In order to assess how presidents have responded to the environment, we will focus our attention on the modern presidency, beginning with Franklin Roosevelt. Roosevelt was the first modern president to pay considerable attention to the environment. As a result, FDR was accorded the honor by environmentalists of having introduced the "Golden Age of Conservation" to America.[1] Much of his effort for and on behalf of the environment focused on his first term in office (1932–36) and was, not surprisingly, linked to the critical concerns of the Great Depression.

Personal Profile

FDR and the "Golden Age"

When we think of Roosevelt and the environment we are likely to think of Teddy Roosevelt, Franklin Roosevelt's fifth cousin. While "TR" has many environmental successes attached to his name, among modern presidents few can rival Franklin D. Roosevelt's accomplishments. Indeed, it was Franklin Roosevelt who has been applauded for introducing the "Golden Age of Conservation."[1]

Franklin Roosevelt, unlike others, was able to take his passion for forestry and the out-of-doors and use it to confront the disastrous unemployment and depressed economy. This president skillfully devised a program to benefit both the environment and the economy, establishing the Civilian Conservation Corps (CCC) to employ millions of young men desperate for employment to work in the national parks

and forests, building roads and trails and planting trees to help prevent
flooding and soil erosion. For him, conservation was closely tied to
America's values. As he stated: "There is nothing so American as our
national parks. The scenery and wildlife are native. The fundamental
idea behind the parks is native. The parks stand as an outward symbol
of this great human principle."[2]

In looking at various aspects of the Roosevelt presidency, we can
tell the seriousness of his efforts for and on behalf of the environment by looking, first, at those who were in his administration. The
most important of those sympathetic to conservation included Henry
Wallace, FDR's secretary of agriculture and, later, vice president, and
Harold Le Clair Ickes, Roosevelt's secretary of the interior from 1933
until 1946. These men were FDR's most important and closest advisers
when it came to conservation.

Equally important, of course, was the growth in government
focus on resolving environmental concerns. New agencies created by
the Roosevelt administration to respond to some of these problems
included the CCC, as well as the Soil Conservation Service, the Soil
Erosion Service, the National Resources Board, the Works Progress
Administration, and the Division of Grazing.

Beyond the appearance of this administration, Roosevelt's skill at
directing Congress and its leadership allowed him to become an effective
negotiator and organizer of coalition support. In addition, Roosevelt
devised a healthy budget for conservation projects, at 20.9 percent of
the general budget in 1935.[3]

FDR showed great strength in the weakest of presidential roles,
namely, as opinion/party leader. Roosevelt knew how to communicate
with the public and sell his conservation program to Congress.

Overall, FDR was a strong environmental leader and an effective
"environmental president." He tended to maximize his own personal
resources and skills in making conservation important as a White House
value.

Notes

1. Richard Lowitt, "Conservation, Policy On," in *Encyclopedia of the American Presidency*, ed. Leonard W. Levy and Louis Fisher, 4 vols. (New York: Simon and Schuster, 1994), 1: 289.

2. "Radio Address Delivered at Two Medicine Chalet" (August 5, 1934) in *Public Papers and Addresses of Franklin D. Roosevelt: The People Approve*, compiled by Samuel I. Rosenman, 5 vols. (New York: Random House, 1938), 3: 359.

3. Byron W. Daynes, William D. Pederson, and Michael P. Riccards, *The New Deal and Public Policy* (New York: St. Martin's, 1998), 116.

Looking at the years since Franklin Roosevelt, the period from 1945 to 1974 was a most productive period as far as protecting the environment and conserving natural resources.[2] During these years, securing the coastal zones, preserving sea mammals, as well as restricting ocean dumping, combating water pollution, and limiting pesticides became priorities for a number of presidents and congresspersons.

Roles of the Modern President

In considering how presidents have dealt with the environment, we will analyze the modern presidency by examining a president's five major roles. The theoretical framework we will adopt revolves around the role approach first developed by Raymond Tatalovich and Byron W. Daynes in *Presidential Power in the United States*.[3]

The five presidential roles we will examine consist of (1) Commander-in-Chief, the only presidential role specifically listed in Article II of the Constitution, and a role that designates the president as the nation's highest ranking military leader; (2) Chief Diplomat, a role that defines a president's position as spokesperson to and negotiator with other nations; since both of these roles call for the president to deal with foreign affairs, we will combine them in looking at how the president deals with the environment in an international setting.

The third role we will consider is that of (3) Chief Executive, a role associated with governing, which involves a president interacting with the bureaucracy, with his or her administrative staff,[4] cabinet, and domestic policymaking in general. The president's fourth role, that of (4) Legislative Leader, indicates the president's important relationship with Congress. Finally, we will take account of the president as (5) Opinion/Party Leader, a combined role linking the president to the public through his political party and in the arena of public opinion.

The influence of each of these roles depends on formal authority, and such political resources as: (1) the ability of a president to make decisions; (2) the public's potential to disapprove of presidential actions; (3) a president's individual expertise in exercising the role; and (4) conditions of crisis that may enhance and expand the power of the president in any one of these roles.

Based on these variables and the political resources normally attending these roles, the five presidential roles can be distributed along a power continuum, as seen in Figure 5.1, from the two strongest roles—the commander-in-chief and chief diplomat—to the weaker two roles—legislative leader and opinion\party leader—with the chief executive role falling in the middle. While there are exceptions to this distribution based on an individual president's ability to reach out to the public, one can normally expect these roles to distribute themselves along the continuum as described.

| COMMANDER-IN-CHIEF | CHIEF DIPLOMAT | CHIEF EXECUTIVE | LEGISLATIVE LEADER | OPINION/PARTY LEADER |

Figure 5.1. Presidential Roles (*Source:* Based on Byron W. Daynes, Raymond Tatalovich, and Dennis L. Soden, *To Govern a Nation: Presidential Power and Politics* [New York: St. Martin's, 1998], Chapter 1, "Presidential Roles, Power, and Policy.")

The President as an Environmental Opinion/Party Leader

Beginning with an assessment of what is considered the weakest of these roles, namely, the opinion/party leader, a president is left on his own to develop ways to reach out to the public since the role has no statutory authority. In this role the president attempts to mobilize public support both for the policies and programs of the administration as well as for the party. Presidents have employed a variety of techniques as opinion/party leader in the effort to influence public opinion, such as news conferences, major and minor policy speeches, and appearances before partisan and nonpartisan groups. Yet one of the major checks on a president's using this role has been the disinterest frequently shown by the public in environmental affairs when compared to other issues. This was shown in a March 2013 Gallup poll that asked 1,022 adults the following question: "Name the most important problem facing the United States today." Fifty-seven percent mentioned an economic issue, while only 2 percent named the environment/pollution as the most important problem.[5] This has been consistent over the years. Only on rare occasions has the populace selected the environment over other issues as most important.

    Those presidents who have considered the environment a primary priority have often made reference to it in their most important address, namely, their State of the Union message. If one finds reference to the environment in this speech, one can be assured that it is a high priority for the president. On several occasions presidents Franklin Roosevelt, Harry Truman and Dwight Eisenhower used their State of the Union messages to emphasize their concern regarding natural resources, stressing the need to make "wise use" of these resources. Early in Richard Nixon's presidency, he declared the 1970s as the "decade of the environment." To show how serious he was about this, he reserved a portion of his 1970 State of the Union address to frame its importance suggesting that "restoring nature to its natural state is a cause beyond party and beyond factions."[6] It may have seemed ironic that Richard Nixon articulated such support since the president had noted, early in his first term, that he had never really been interested in environmental issues. As he told John Ehrlichman, one of his closest aides, in a recorded conversation, he thought environmentalists were all "overrated," serving only the "privileged." The environment as an issue, Nixon continued, "was just 'crap,' and for 'clowns,' and 'the rich and [Supreme Court Justice William O.] Douglas.'"[7] Yet Nixon was a consummate enough politician to recognize a popular issue when he saw it. This was an unusual period, when the public, the Congress, and both parties were in agreement in support of the president's environmental focus. Nixon determined in his first term that his focus on the environment would be on clean water, constraining air pollution, and on finding new ways to increase the number of national parks and expand open space.[8] During the Nixon presidency more than half (53 percent) of the public indicated that environmental quality was the most important problem facing the nation.[9]

But it was George H. W. Bush who announced during his election campaign for president that he wanted to be known as the "environmental president." Such an announcement immediately separated him from Ronald Reagan, whom he had served as vice president. President Reagan, as it turned out, preferred to support development and regulatory relief for business, which tended to undermine environmental protection. Despite Bush's announcement, however, many environmentalists were not convinced by his rhetoric, believing that he had reversed his own position too often on environmental issues as vice president to now be credible as an "environmental" president. This was a prediction that proved to be true late in the Bush presidency.

Rather than offering support for George H. W. Bush, environmentalists enthusiastically looked to Bill Clinton as their "great green hope" after enduring twelve years of the Reagan and Bush presidencies. Yet Clinton, as opinion/party leader, actually emphasized the environment in fewer speeches than any of his Democratic or Republican predecessors, although in his last State of the Union message he did make clear that the environment had been and would continue to be an important issue of concern during his presidency. Moreover, it was in his 1994 Earth Day speech that Clinton made his public and party commitment to environmental protection quite clear by citing the words of President John Kennedy, who had declared, "It is our task in our time and in our generation to hand down, undiminished to those who come after us, as was handed down to us by those who came before, the natural wealth and beauty which is ours."[10]

In President Barack Obama's 2014 State of the Union message, he made several references to the environment when he suggested that he and his administration would strengthen the "protection of our air, our water, our communities. And while we're at it, I'll use my authority to protect more of our pristine federal lands for future generations." He then stated that he had directed his administration to "work with states, utilities and others to set new standards on the amount of carbon pollution our power plants are allowed to dump into the air." Finally he made reference to climate change, suggesting that "the debate is settled. Climate change is a fact. And when our children's children look us in the eye and ask if we did all we could to leave them a safer, more stable world, with new sources of energy, I want us to be able to say yes, we did."[11]

Case Study

Leaving a Legacy: Barack Obama and Climate Change

What will American citizens remember most about Barack Obama after he is no longer president of the United States? Will it be the Affordable Care Act, tax reform, or social equality? Obama, entering the latter part of his second term, wants to leave his stamp on the history of the United States. Therefore, he has made the environment a major focus of his administration's later years in office. Secretary of State John Kerry clearly supports the president's focus, because he has compared the threat posed by climate degradation as equal to that of "weapons of mass destruction."[1] He intends that the United States lead the world in environmental reform by first starting at home. " 'When America proves what's possible, other countries are going to come along,' he said."[2] Although this is Obama's vision for his presidency, he may be thwarted by House Republicans unless he can find ways lawfully to act without their approval.

As one might imagine, there is considerable opposition to Obama's environmental initiatives. Said one article, "He is already facing back- lash—not only from Republicans, but also from Democratic lawmakers running for re-election in coal-heavy states."[3] Many of the concerns come from states dependent on energy production from coal. These states and their representatives (who are usually heavily Republican) fear that Obama's new regulations will lose jobs and raise energy costs. The opposition also believes Obama's changes will hurt the overall economy of the United States. However, a report from the United States Chamber of Commerce showed the costs of policy change will not be as high as some coal defenders think.[4] Nonetheless, House Republicans and even some Democrats will still work hard to stop Obama in his venture to curb climate change.

To circumvent the opposition, President Obama intends to use preexisting laws and his own executive authority. "Climate change remains among the few policy items he can push without action from Congress."[5] For example, Obama plans to use laws such as the Clean Air Act to curb the emission of harmful gases into the environment. "The President promised that the United States would reduce its greenhouse gas pollution by 17 percent from its 2005 level by the year 2020 and by 83 percent by the year 2050."[6] By doing this, President Obama does not need Congress in order to act; Congress already gave its approval.

In seeking to assure his legacy, President Obama released a plan with seven goals: (1) assess the impact of climate change; (2) support climate-resilient investments; (3) rebuild the areas damaged by Superstorm Sandy and use the lessons learned from the catastrophe; (4) launch an effort to create sustainable and resilient hospitals; (5) maintain agriculture productivity; (6) provide tools for climate resilience; and (7) reduce the risk of droughts and wildfires. These goals are "aimed at guarding the electrical supply; improving local planning for flood, coastal erosion and storm surges; and better predicting landslide risks as sea levels rise and storms and droughts intensify."[7]

Obama and his administration went to work right away to achieve these goals. The administration has already released its third United States National Climate Assessment to further explain what it intends to do regarding climate change. The administration is also offering grants-in-aid to foster the development of more resilient technologies to protect the environment, and has developed committees for the same purposes. The president is making sure that his efforts to remedy the climate crisis are not seen as his efforts alone, but will be viewed as a way to share information with Americans and the world regarding what the environmental necessities are in responding to the climate crisis. Therefore, everyone can be more informed on how to better react to climate change.[8] Obama is also using his Hurricane Sandy Task Force to learn from that climate disaster so that next time the nation faces a similar threat, the United States will be better prepared.

Other presidents have sought to bring about environmental policy reform, as well. However, Obama appears much stricter when implementing policies. For instance, "Bush advocated 'voluntary means' but President Obama makes it quite clear he will insist on 'mandatory means' with regard to reducing emissions from coal-fired power plants."[9] Another example of Obama encouraging "mandatory means" is in his use of executive orders. On November 1, 2013, President Obama issued

Executive Order 13653—Preparing the United States for the Impacts of Climate Change. In this action, Obama ordered government agencies to liberally share environmental information with the public to assist in "risk-informed decision making." He also hopes this will better prepare the nation for the future.[10] President Obama is taking action against climate change and utilizing every power he possesses to encourage the federal government and society, as a whole, to curb climate change. Some governmental agencies and industries have joined Obama in his climate goals, such as the EPA, while others have put up major resistance to what he is doing, such as the coal and oil industries.

President Obama showed the United States and the world that climate change is his legacy. Many people laud him for his initiative and vigor in pursuing this environmental cause. Some say even more can be done. Said one reporter, "If we're still getting over 30% of our power from coal in 2030, the EPA's plan will be a huge disappointment."[11] Can more be done? One can always say yes, but President Obama is aggressively attacking the issue and is doing what he can to save the environment—leaving a legacy.

Notes

1. Byron Daynes and Glen Sussman. "The Political Response to Climate Change in the USA and Canada," *The USA and Canada* (2014), 14–29.

2. Coral Davenport, "Climate Can't be Deaf to Economic Worries, Obama Warns." *New York Times*, June 26, 2014: http://www.nytimes.com/2014/06/26/us/politics/obama-warns-climate-cant-be-deaf-to-economic-worries.html.

3. Andrew Leach and Luiza Savage, "Obama's Climate Change Fix," *Maclean's*, June 9, 2014, 28–30.

4. John Young, Eugene Robinson, Froma Harrop, Paul Krugman, and Thomas Friedman, "Obama: True Champion of the Planet," *Liberal Opinion Week*, June 18, 2014.

5. Davenport, "Climate Can't be Deaf."

6. Daynes and Sussman, "The Political Response to Climate Change."

7. "Climate Change and President Obama's Action Plan"; http://www.whitehouse.gov/climate-change. Accessed July 21, 2014.

8. "Climate Change and President Obama's Action Plan."

9. Daynes and Sussman, "The Political Response to Climate Change."

10. Barack Obama, "749-Executive Order 13653—Preparing the United States for the Impacts of Climate Change," The American Presidency Project. November 1, 2013. Accessed March 18, 2014.

11. Michael Grunwald, "Obama's Carbon Rules: Why the Numbers in the President's Clean Power Plan Don't Add up," *Time*, June 30, 2014, 26.

Summary

As opinion/party leader, the president lacks many of the resources that accrue to the other presidential roles and must therefore rely on his own personal skills to reach out to the public. One might say, then, that the opinion/party leader role is important but weak. It is important because a president can influence environmental policy in this role, and can persuade public sentiment for or against environmental protection. It is weak, however, because a president faces a number of constraints including a divided political system, without the authority and resources of other roles to easily compensate for this.[12]

The President as an Environmental Legislative Leader

For most presidents, the legislative leader role grants access to moderate political resources. It is, however, a role in which Congress has a substantial advantage if not the dominant influence;[13] Congress can either support presidential action, shape a president's agenda, or come out in direct opposition to the president's efforts. A president who is a successful legislative leader must often rely on external political resources, creativity, and political persuasion to facilitate an administration's environmental focus. A president must assertively use all of his individual skills as a negotiator and persuader, since there is little real authority attending this presidential role. We have had several outstanding examples of such presidents. Franklin Roosevelt established himself in 1932–33 as an effective legislative leader in the now-classic "first hundred days," during which time he fostered the passage of fourteen major pieces of environmental legislation, including legislation establishing the Civilian Conservation Corps and the Tennessee Valley Authority.[14] But it was Lyndon Johnson who proved to be an even more effective legislative leader, based on his many years of experience as Senate majority leader. While president, he encouraged the passage of several important pieces of environmental legislation, including the Clean Air Act, the Wilderness Act, as well as legislation focused on water resources, land and water conservation, solid waste, highway beautification, natural gas, noise abatement, rivers, and trails.[15] Dennis L. Soden and Brent S. Steel, in their examination of Johnson, in fact, list eighteen major environmental bills that were passed during the Johnson years, most of which were signed by the president during his first term in office.[16]

Richard Nixon, assured that the 1970s would be the decade for environmental awareness, used this role to support passage of the National

Environmental Protection Act, to create the Environmental Protection Agency, and to encourage the passage of more pieces of environmental legislation than were passed in any other modern-day Congress.[17] Of the twenty-five pieces of environmental legislation that became law under Nixon, all but four were passed during his first term.[18]

Environmentalism, and the energy crisis in particular, took much of Jimmy Carter's attention domestically. Of the Carter environmental bills that were most important during these years, the Alaska National Interest Lands Conservation Act of 1980 (PS 96-487) was paramount. It was Carter's most controversial effort, as well, if for no other reason than that the 103 million acres of land that were to be set aside by Congress and the president gave rise to a great deal of concern on the part of the Alaska delegation in the Congress. As Cecil Andrus and Joel Connelly put it: "Carter managed to leave behind a legacy of volcanic craters, alpine lakes, ancient forests and tundra, and federal land managers who weren't devoted only to drilling, digging up and cutting down the great resources of America's forty-ninth state."[19]

Ronald Reagan has posed somewhat of a puzzle for students of the presidency and the environment. He came into office with the desire to cut back on the environmental advances that had been made over the preceding years. And Ronald Reagan did not disappoint those favoring development over the environment, for he vetoed more environmental legislation than he supported. For example, he opposed the Superfund Amendments and Reauthorization Act of 1986 (PL 99-499), the Nuclear Waste Policy Amendments (PL 100-203) of 1987, as well as the Clean Water Act Amendments of 1989 (PL 100-203).[20]

Bill Clinton's record was mixed as well. He was quite successful in reaching both Democrats and Republicans in the first year of the 103rd Congress, encouraging them to support his environmental priorities. But Clinton could not strike a consensus with Republicans in the 104th Congress, once both the House and Senate became Republican. The 104th Congress, in the view of the League of Conservation Voters, had perhaps "the worst environmental voting record of any Congress in the past 25 years."[21] Clinton's environmental record was much more positive in his second term.

Barack Obama has made an unsuccessful effort to encourage Congress to pass bipartisan environmental legislation. It has been the House Republicans that have prevented him from being as effective as he has wanted to be. In commenting on the beginning of the current 113th Congress, the League of Conservation Voters had this to say:

Just as they did in the 112th Congress, the House seemingly left no issue untouched during the first session of the 113th Congress. The attacks included efforts to: roll back cornerstone environmental laws like the Clean Air Act, the Clean Water Act and the National Environmental Policy Act; legislatively approve the risky Keystone XL tar sands pipeline and increase harmful drilling and fracking across the country; decimate protections for our forests and other public lands; continue subsidizing dirty fossil fuels while cutting funding for renewable energy and energy efficiency; and deny the costs of carbon pollution despite the fact that they are already all too apparent.[22]

This Congress seems to rival the one faced by President Clinton in terms of their resistance to searching for environmental solutions. Moreover, House Republicans have attempted to dismantle environmental advances that have been made by other presidents. The good news for environmentalists, as the league acknowledges, has been the efforts of the Senate and the Obama Administration to support environmental gains. As the league observes, the Senate and administration have "blocked the vast majority of House-passed attacks on the environment and public health."[23]

Powers Available to the President as Legislative Leader

These powers include a president's Article II veto power, which gives the president his greatest leverage in the legislative process since so few regular vetoes are ever overridden by a two-thirds vote in both houses of Congress.[24] Presidents, with a focus on the environment, have relied on the veto power primarily to protect environmental gains they have secured as legislative leaders as did Bill Clinton, who on October 22, 1999, vetoed an Interior Department budget bill passed by Congress because it did not fully fund his environmental program, which included his Land's Legacy program, his climate change proposal, his clean water effort, and the environmental assistance he was giving Native Americans.[25] Presidents who have vetoed environmental legislation included George W. Bush who vetoed the Water Resources Development Act of 2007 because, in his judgment, "[t]his bill lacks fiscal discipline. I fully support funding for water resources projects that will yield high economic and environmental returns to the Nation. . . . However, this authorization bill makes promises to local communities that the Congress does not have a track record of keeping."[26]

Other presidents have found it was sufficient to merely threaten the use of the veto power rather than to actually use it in order to alter and change the focus of legislation to make it more to their liking. Charles M. Cameron argued in 2010 that the veto threat happens most frequently during divided party government and that it has been "remarkably efficacious in extracting concessions" from the Congress.[27] Jimmy Carter was one president who made use of veto threats. One example was on a public works bill (HR 12928) that Carter opposed. It involved six water projects among the eighteen projects in 1977 that he wanted killed. His indication that he would veto this was sufficient to remove the objectionable measures from the bill to appease the president.[28]

THE BUDGET

A successful environmental program requires funding. One can tell how important the environment was to Franklin Roosevelt's administration by examining the budgetary allotments he set aside for the environment. He devoted as much as 20.8 percent of his budget in 1935 to environmental projects despite needing also to respond to the Great Depression.[29] Other presidents have also shown how crucial money is in supporting environmental policy. Richard Nixon, for example, asked Congressional leaders in his Annual Budget Message for Fiscal year 1972 for 2.4 billion dollars more than Congress was willing to provide for environmental proposals.[30] Despite Congress's refusal, Nixon the next year asked Congress for five billion dollars for the environment, but this time he requested that the money should be allotted to the states to assist them in their environmental programs.[31] This request, not surprisingly, was approved.

Ronald Reagan found he could just as easily use budgetary funds to starve environmental legislation as to support it. In 1985 Reagan cut funding of a Soil and Water conservation bill in order to "avoid the specter of higher interest rates, choked-off investment, renewed recession, and rising unemployment."[32]

BILL SIGNING

Some presidents have found their environmental priorities have attracted greater visibility through bill signing ceremonies. George H. W. Bush, for example, indicated a number of times that clean air was a high priority of his, particularly in trying to improve the air in polluted urban areas. Bush, in addition to signing the Clean Air Act of 1990, also issued a "mission" statement suggesting that the passing of the law had been the culmination of a year-long crusade on his administration's part to fashion this legislation.

When it appeared, the public needed another reminder that Bush had labeled himself an "environmental president," and the president illustrated the importance of this environmental agreement by signing it on September 18, 1991, on the rim of the Grand Canyon. This still did not convince everyone that Bush was a sincerely committed environmentalist and so the next year when he signed the Los Padres Condor Range and River Protection Act in June 1992, he did so in Washington, D.C.[33]

Summary

There has been significant environmental legislation, as Glen Sussman and Mark Kelso argued, that every president since Kennedy has at least signed. Those who were most active in encouraging the passage of pro-environmental legislation that added important ingredients to the environmental movement include Franklin Roosevelt, Lyndon Johnson, Richard Nixon and Jimmy Carter.[34]

As legislative leaders, Richard Nixon and George H. W. Bush—and Bill Clinton, to a limited extent in his first term—also employed creative techniques to lead Congress into becoming a positive part of the environmental movement. These presidents were all quite active in their involvement in protecting the environment. President Obama hoped to be a major contributor in support of the environment, as well, but opposition from Republicans in Congress have frustrated his intentions. Yet if one considers all of the votes in this divided Congress, Obama's success rate is greater in passing legislation he has supported than any other president since Eisenhower. For example, in 2012 Obama saw 54 percent of those bills he supported pass compared to George W. Bush in 2007, when Bush saw only 38.3 percent of the bills he supported pass. Bill Clinton, in 1995—his least successful year—saw only 36.2 percent of the bills he supported pass the divided Congress. It has been on the environmental bills where President Obama has been least successful.[35] For example, on March 6, 2014, the U.S. House of Representatives passed the "Electricity Security and Affordability Act" (HR 3826) by a vote of 229–183, a bill that would allow unlimited carbon pollution from power plants, which clearly undermines the Clean Air laws. The president opposed this bill in the House, but he was unsuccessful in preventing its passage. The Senate has yet to take up the bill.[36]

The President as an Environmental Chief Executive

Because the framers of the Constitution determined that the chief executive should not share power with any advisory body, the chief executive role may seem quite powerful. Yet the Constitution fragments control over the bureaucracy, often making it difficult for the president to exert meaningful control over the executive branch.

When it comes to influencing policy, however, the president as chief executive does have at least two effective resources, namely, the creation of office structures and in the staffing of government.

OFFICE STRUCTURES

Franklin Roosevelt took full advantage of this power, fulfilling both his passion for conservation and his desire to put people back into the workforce. Approved in 1933 by Congress, the Civilian Conservation Corps (CCC) became the first of Roosevelt's New Deal Agencies. Not only was the CCC effective in putting young men back to work during the Great Depression, it was also Roosevelt's most important agency for protecting the environment. Many men were employed to enhance national parks and forests by building roads, establishing visitor pathways, and making other improvements.[37]

The Environmental Protection Agency (EPA), was established in 1970 by Richard Nixon as an independent executive agency to protect both public health and the environment. As of FY 2014, it had an annual budget of approximately $8.2 billion and a staff of 15,913 persons.[38] President Obama has used the EPA as an effective instrument in circumventing congressional opposition to his environmental policies.[39]

Bill Clinton considered the EPA to be so important that he attempted to elevate it to cabinet status.[40] Unfortunately for environmentalists, this proposal failed to pass the Congress.[41]

Another important agency that has helped to shape environmental policy is the Federal Emergency Management Agency (FEMA), which was established during Jimmy Carter's administration in 1979 to protect property in the event of national emergencies or disasters. As climate change has brought about more frequent weather-related natural catastrophes, FEMA has been called on many times in relief of disaster victims.

At times, reorganization may work against a president and his priorities. This happened during Clinton's years in office. Clinton and Gore's "reinvention" of government, as the policy was called, allowed the Clinton administration to reorganize and reduce the size of the federal government, making government less expensive and more efficient.[42] In carrying out these reforms, Clinton saw some support for environmental policies disappear. There was a 12 percent reduction in the federal workforce, including in some of the agencies that dealt directly with environmental policy.

PRESIDENTIAL APPOINTMENTS

While restructuring has proven to be an effective tool for the chief executive, the president's responsibility to staff the government has been even more important in responding to critical needs.

The cabinet is the oldest advisory body in the executive branch, but the cabinet as a whole has rarely been important in terms of strengthening a president's social agenda. Individual cabinet secretaries, however, have been helpful in this regard. During his first term, Franklin Roosevelt effectively used staffing appointments and the creation of new offices to advance his conservationist agenda. His most important appointments included Harold Ickes, his secretary of interior; and Henry Wallace, his secretary of agriculture, who were FDR's closest, and at times, most impassioned advisors on conservation. Many conservationists also served in informal advisory roles in Roosevelt's "kitchen cabinet."

One reason for Lyndon Johnson's success on environmental protection was the work of his secretary of the interior, Stewart Udall, a Kennedy hold-over appointee. Udall was strongly supported by environmentalists. He was an enthusiastic backer of conservation and national park expansion, and worked to improve the quality of people's experiences in the parks. Through Udall's leadership at the Interior Department, as well as the strong public support for Johnson's environmental program, LBJ was effective in addressing environmental issues.

One could immediately see the change in support for environmental protection when Ronald Reagan came into office. Reagan slowed environmentalism through the appointments he made to his administration. As Richard Nathan described it, Reagan's "administrative presidency" worked to reverse years of bipartisan support for the environment in the Congress.[43] President Reagan's campaign against federal environmental laws led him to appoint Anne Burford as the head of the Environmental Protection Agency. He encouraged her to limit environmental funding, as well as to deregulate and restrict the enforcement of federal environmental rules. Reagan used his authority to encourage the EPA to be more supportive of business interests. He further showed his antagonism for federal involvement with the environment in his appointment of James Watt, his first secretary of interior, and one of the least supportive interior secretaries of environmental protection ever appointed. President Reagan encouraged Watt to limit funding for the Department of the Interior and cut back on the enforcement of federal rules governing the environment, both of which occurred.[44]

In contrast to Reagan, Bill Clinton's pro-environment appointees were numerous. Mark Dowie estimated that Clinton hired about "two dozen environmentalists."[45] Clinton's placement of these environmentalists in such unlikely institutions as the State Department,[46] the National Security Council, and the Office of Management and Budget is even more impressive.[47] Vice President Al Gore, Secretary of the Interior Bruce Babbitt, and Carol Browner, the administrator of the Environmental Protection Agency (EPA), were three of Clinton's most important environmentalist appointments; each brought strong pro-environmental credentials from their previous positions.

In examining the appointments of George W. Bush, it becomes clear that environmental protection was not an issue he was interested in supporting. As professors Sussman and Daynes pointed out:

> The interests represented by George W. Bush's appointments included the fossil fuel industry, mining, ranching and timber. As heads of agencies, Bush's appointees were in a position to use their power to appoint lower-level officials and staff that would most likely share the Bush environmental philosophy.[48]

As Bush's vice president, Dick Cheney, a petroleum entrepreneur and former congressman from Wyoming, opposed the enforcement of environmental protection laws. Other anti-environmental appointees to the Bush administration included Thomas Sansonetti, and Jeffrey Holmstead. Sansonetti, before being appointed Assistant Attorney General for Environment and Natural Resources, had been a lobbyist for coal mining operations and other industries that sought access to public lands. Holmstead, who was appointed as assistant Environmental Protection Agency administrator, had represented electric utilities as a lobbyist with Latham and Watkins, which had fought against air pollution restrictions. He had also been a member of the Farm Bureau, which had opposed pesticide controls. Bush chose no appointees with impressive pro-environmental credentials.

In stark contrast to Bush's appointees, most of those appointed by Barack Obama had impressive environmental experience. Lisa Jackson, Obama's first Environmental Protection Agency administrator, had previously worked for sixteen years for the New Jersey Department of Environmental Protection, and Gina McCarthy, Jackson's successor, had been an environmental advisor to five Massachusetts governors, and has had strong support from environmental groups. Carol Browner, the former director of the White House Office of Energy and Climate Change Policy, had headed the Clinton EPA, but served only a short time during Obama's first term. Three other appointments worth mentioning include Ernest Moniz, secretary of energy, who had been director of energy and the environment at MIT; Mike Boots, chief of staff of the Council on Environmental Quality (CEQ), who has worked with U.S. and European governments and businesses on a broad-ranging number of environmental policies and assists the president in his Climate Action Plan; and John Kerry, as secretary of state, who had a twenty-year record of combating climate change during his time in the Senate. Even President Obama's secretary of the interior, Sally Jewell, who had worked as a petroleum engineer at Mobil Oil Corporation, won the 2009 Rachel Carson Award for Environmental Conservation from the Audubon Society[49] and served as a board member of the National Parks Conservation Association.

Powers Used by the Chief Executive to Advance Environmental Priorities

Many presidents have used executive orders to affect change in environmental policy. By relying on this power, a president is given some independence from Congress in structuring his programs. Franklin Roosevelt used executive orders to advance conservation and environmental policy in 1934 and to strengthen his influence over the environment by redesignating public lands for conservation purposes.[50] Other presidents who were not as supportive of environmental advancement used executive orders to change environmental policy. Ronald Reagan established Executive Order 12291, which enabled the Office of Management and Budget to prevent all regulatory action unless the office deemed that its "potential benefits to society . . . outweigh the potential costs."[51]

Bill Clinton introduced several new environmental programs through the use of executive orders including his 1993 clean air program and his 1999 Clean Water Action Plan.[52] Of George W. Bush's nineteen executive agreements dealing with the environment, five of them weakened environmental protections. Many of Bush's apparently pro-conservation executive orders were misleading, for example, Executive Order 13443 (August 16, 2007), Facilitation of Hunting Heritage and Wildlife Conservation, was issued for the purpose of conserving wildlife for hunting purposes. Executive Order 13474 (September 26, 2008) weakened Executive Order 12962, which had ensured recreational fishing in national wildlife refuges, national parks, national monuments, national marine sanctuaries, and marine protected areas.[53] Executive orders 13352 and 13366 did not get sufficient funding, so they were of no consequence. Executive Order 13287, to "Preserve America," served the primary purpose of encouraging tourism and was not aimed at protecting the environment. One executive order that could clearly have lent itself to abuse of the environment was Executive Order 13406.[54] This particular executive order gave government officials the power to take any land that would benefit "the general public." This could easily have opened the way for environmental abuse, given the lack of support in this administration for environmental protection.

Barack Obama has issued a total of fifteen executive orders dealing with the environment, all of which have been supportive of environmentalism. Peniel E. Joseph has argued that Obama's executive orders have been a way for the president to accomplish his environmental agenda in light of the resistance to his environmental programs from the Republican-controlled House of Representatives.[55] All of Obama's executive orders were much longer and more complex than Bush's executive orders. On October 9, 2000, for example, President Obama emphasized his commitment to combating global climate change by issuing Executive Order 13514, entitled "Federal Leadership in Environmental, Energy, and Economic Performance," for the purpose of making the "reduction of greenhouse gas emissions a priority for Federal agencies."[56] In 2013, Obama, acknowledging the reality that he was unlikely to convince the Republican-dominated House of Representatives to support his environmental agenda, chose to issue three executive orders, which forcefully illustrated his seriousness about protecting the environment. The 2013 executive orders included Executive Order 13648 (July 1, 2013), Combating Wildlife Trafficking; Executive Order 13650 (August 1, 2013), Improving Chemical Facility Safety and Security; and Executive Order 13653 (November 1, 2013), Preparing the United States for the Impacts of Climate Change. This last executive order gives the president and a Task Force the authority to modernize federal programs to "support climate resilient investment," manage lands and waters for "climate preparedness and resilience," "provide information, data, and tools for climate change preparedness and resilience," and have each federal agency come up with a plan for "climate change related risk."[57] President Obama, in these executive orders, made clear his support for confronting the pressing issue of climate change.

Summary

The chief executive role offers presidents a moderately strong platform from which to strengthen environmental policy. Important progress can be made in facilitating an environmental agenda through both the creation of new federal agencies and the reorganization of existing agencies. As important as creating new administrative structures are the president's opportunities to make staffing appointments and to issue executive orders. The appointment of individuals to key governmental positions allows the president to shape environmental policies throughout various agencies. Presidential appointments and executive orders give the chief executive great influence in either protecting the environment, as did President Roosevelt, or in weakening environmental regulations, as did Presidents Reagan and George W. Bush.

The President as Environmental Commander-In-Chief/Chief Diplomat

The two roles that focus a president's attention on foreign affairs are the commander-in-chief and chief diplomat roles. These are a president's strongest roles in terms of the authority and resources at his disposal. The power of the commander-in-chief is impressive because decision making tends to be centralized, and major decisions may involve only the president and a select number of key advisers.

A president's role as chief diplomat is almost as powerful. Constitutionally, authority over foreign affairs is shared between the president and the Congress, but custom, judicial decisions, and statutory law have favored executive dominance by the president in foreign affairs. Despite the fact that these are the president's most powerful roles, in the past they have seldom been used to influence environmental policy. Yet globalization has drawn the world's nations closer together and made environmental policy a major international issue. In confronting environmental issues with international consequences, the president's constitutional roles put him in a strong decision-making position.

Although environmentalism has not been a part of every president's agenda, it will be more difficult in the future to disregard international environmental difficulties such as dwindling natural resources and the deterioration of air and water quality around the globe. Table 5.1 shows the proportion of environmental agreements concluded by American presidents in relation to the total number of international agreements.

Table 5.1. Number of International Agreements (by President) Dealing with the Environment as a Percentage of all Agreements, 1949–1996

President	International Environmental Agreements (N)	Percent of All Agreements (%)
Truman	96	11.3
Eisenhower	275	14.3
Kennedy	117	14.1
Johnson	247	22.2
Nixon	248	19.1
Ford	165	21.3
Carter	312	26.7
Reagan	350	28.0
Bush, G. H. W.	186	18.4
Clinton	426	24.9
Bush, G. W.	350	16.1
Obama	172	20.8

Source: This Table is adapted from material in Lyn Ragsdale, *Vital Statistics on the Presidency: Washington to Clinton* (Washington, DC: Congressional Quarterly, 1996), 318–21. The more current data on Bill Clinton came from Lyn Ragsdale, *Vital Statistics on the Presidency*, revised ed. (Washington, DC: Congressional Quarterly, 1998), 329. Information on George W. Bush and Barack Obama came from Igor Kavass, *The Current Treaty Index*, 25th edition (1999), 29th edition (2001), 37th edition (2005), 45th edition (2009), and the 53rd edition (Buffalo: W. S. Hein, 2013).

Clearly, environmental agreements constitute a small proportion of all international agreements, ranging from a high of 28 percent to a low of 11.3 percent. Yet all presidents from Franklin Roosevelt through Barack Obama have been responsible for some agreements, either on a bilateral or multi-lateral basis. Some significant international agreements address "new" global environmental threats. One example was the Montreal Protocol of 1987, an agreement that required signers of the document to agree to work toward substantive reductions in the production of chlorofluorocarbons (CFCs). Ronald Reagan surprised environmentalists, who as a whole had been among his most aggressive critics, by signing the Montreal Protocol on Ozone Depletion. At the 1988 signing, President Reagan touted the protocol as a "model of cooperation," and a "product of the recognition and international consensus that ozone depletion is a global problem."[58]

In 1991, George H. W. Bush signed an air quality agreement with Canada. This was an important step in solving the shared problems of acid rain and poor air quality.[59] Yet at an Earth Summit in Rio, Bush disappointed environmentalists by agreeing to sign the 1992 United Nations Convention on Climate Change only after its standards had been significantly weakened. The agreement, which asked signees to cooperate in sharing environmental technologies and in sharing the costs of maintaining the environment, was undercut when Bush used his influence to ensure that it excluded placing specific limits on emission levels as well as setting any binding timetables for reducing emissions.[60]

President Obama has been quite active in involving himself in international agreements. In 2009, for example, he and Prime Minister Stephen Harper of Canada committed themselves to work together to develop clean energy science and technologies. Obama stated that a result of their dialogue would be to "advance carbon reduction technologies,"[61] and in 2013 the president, along with the leadership of China and the other G-20 nations, agreed to work through multilateral approaches using the institutions of the Montreal Protocol to "phase down the production and consumption of HFCs," while continuing to use the accounting and reporting of those emissions under the authority of the United Nations Framework Convention on Climate Change (UNFCCC) and provisions specified in the Kyoto Protocol.

In July 2014, the United States and thirteen other members of the World Trade Association began negotiating a reduction in tariffs on environmental goods to assist developing countries in adopting environmentally friendly technologies. The governments involved included Australia, Canada, China, Costa Rica, Hong Kong, Japan, South Korea, New Zealand, Norway, Singapore, and Switzerland. These governments contribute some 86 percent of the environmental goods trade.[62] The twenty-eight member countries of the European Union (EU) also gave support to the agreement.

President Obama, representing the United States in association with Canada and Mexico, also proposed an amendment to the Montreal Protocol that would "phase-down consumption of and production of hydrofluorocarbons (HFCs)." In November 2014 this was formally discussed to determine if it will become a formal part of the protocol.[63]

Summary

The environment has become increasingly important in foreign affairs as the president has become more engaged both in environmental diplomacy and in addressing national security issues related to the environment. Almost every president from FDR through Obama has signed important bilateral and/or multilateral agreements that have affected the environment.

As Woodrow Wilson recognized, the president not only holds an important position in American politics, he is also a central figure in the international community. Presidents now acknowledge that the activities of individual nations have environmental consequences that may be felt on a regional and international scale. The problem of acid rain, addressed during the Reagan administration, is a good example of how the negligence of one nation (the United States) can encroach on the quality of life in another country (Canada). The international debates over stratospheric ozone deple- tion and global warming confirm the interconnections among countries. The prospect of war over access scarce resources such as water or oil may become more prominent among the challenges of the twenty-first century.

Conclusion

Not all presidents have responded in the same way to environmental issues. Some have actively supported environmental protection, while others have been aggressively opposed to them, and still others have only symbolically responded to the challenges posed by the issues, as Table 5.2 indicates. Among the early modern presidents, FDR was one of the most supportive

Table 5.2. Presidential Types Based on their Approach to Environmental Policy

Activist	Symbolic
F. Roosevelt	H. Truman
L. Johnson	D. Eisenhower
R. Nixon	J. Kennedy
J. Carter	G. Ford
R. Reagan[1]	G. H.W. Bush
G. W. Bush	
B. Clinton[2]	
B. Obama[3]	

1. Ronald Reagan was an "activist" in opposition to environmental policy.
2. Bill Clinton had a mixed record and could fit in between the Activist and Symbolic.
3. Barack Obama was prevented from taking as many pro-environmental actions as he wished.

of environmental issues. Of the later modern presidents, Richard Nixon proved to be most successful in promoting environmental priorities, partly because he had the support of the Congress and the public that many of the other presidents lacked.

All presidents have shaped environmental policy in one way or another. Nixon's successors—Carter, Reagan, George H. W. Bush, Clinton, George W. Bush, and Barack Obama—used their powers to shape environmental policy, though to a more limited degree than he.

Ronald Reagan was an activist, but used his power as chief executive to reverse environmental progress, choosing to support economic development at the expense of environmental preservation. In Norman Vig and Michael Kraft's words: "The 'environmental decade' came to an abrupt halt with Reagan's landslide victory in 1980."[64] Much of what Reagan did to undercut the environment, as these researchers suggested, necessitated bypassing the Congress and relying, instead, on his role as chief executive to frustrate environmental advancement through the use of staffing and appointments. Reagan appointed such anti-environmental figures as Anne Burford, who headed the Environmental Protection Agency, and James Watt, who served as secretary of the interior. As well, Reagan severely reduced the budgets and personnel of environmental agencies.[65]

While we have focused primarily on those presidents, among the fourteen modern presidents who were most active in responding to the environment, there have been several who could only be said to offer "symbolic" responses if not ignoring the environment altogether. Those who would fit this category, as Table 5.2 points out, include Harry Truman, Dwight Eisenhower, John Kennedy, and Gerald Ford.

While environmental concerns are never going to dominate the agenda of all presidents, the critical nature of environmental issues such as global climate change will continue to increase the pressure on future American presidents to exert leadership and confront environmental concerns.

Websites

*President and Vice President: www.whitehouse.gov

*Presidential Research: http://www.ibiblio.org/lia/president/President-directory.html

---

*changed

*Executive Office: www.whitehouse.gov/administration/eop

*Cabinet: www.whitehouse.gov/administration/cabinet

Presidential Libraries

*Franklin D. Roosevelt: www.fdrlibrary.marist.edu

Harry S. Truman: www.trumanlibrary.org

*Dwight D. Eisenhower: www.eisenhower.archives.gov

*John F. Kennedy: www.jfklibrary.org

*Lyndon Johnson: www.lbjlibrary.org

*Richard Nixon: www.nixonlibrary.gov

*Gerald Ford: www.fordlibrarymuseum.gov

*Jimmy Carter: www.jimmycarterlibrary.gov

*Ronald Reagan: www.reaganfoundation.org

*George H.W. Bush: bushlibrary.tamu.edu

*Bill Clinton: www.clintonlibrary.gov

*George W. Bush: www.georgewbushlibrary.smu.edu

*Barack Obama: www.barackobamapresidentiallibrary.com

Other Possible Sites

COMPILATION OF PRESIDENTIAL DOCUMENTS

www.gpo.gov/fdsys/browse/collection.action?collectionCode=CPD

PUBLIC PAPERS OF THE PRESIDENT

www.archives.gov/federal-register/publications/presidential-papers.html

MAJOR PRESIDENTIAL SPEECHES

http://www.presidentialrhetoric.com/index.html

---

*changed

6

Executive Agencies and the Environment

As most American citizens know, the Constitution mandated a government comprised of three branches: the legislative (to create laws), the executive (to enforce the laws), and the judicial (to interpret the laws). Yet few of us are familiar with what many call the fourth branch of government: the array of administrative and regulatory agencies. While the Constitution makes no mention of this fourth branch, there are currently over four hundred agencies and more than two million federal civilian employees in 350 different occupations.[1] At the state and local level, far more administrative agencies and public employees carry out the work of government. To help you understand environmental politics and policy, this chapter concentrates on the roles and functions of these administrative agencies.

The fourth branch of government is sometimes referred to as the bureaucracy. While the popular use of that term is often negative, focusing on such unfavorable characteristics as impersonality, sluggishness, and rigidity, more positive characterizations include predictability, neutrality, deliberateness, and ability to mobilize to complete tasks.[2] Actually, dividing the word into two parts—*bureau* ("office" in French) and *cracy* ("form of rule" in Greek)—helps us appreciate the meaning of the term. Classic writings spell out the crucial elements of bureaucracy's inner structure, beyond the notion of "government by offices and desks."[3] Six characteristics stand out: division of labor (fixed jurisdictional areas), hierarchy (clear superior-subordinate relations), written documentation (as opposed to oral understandings), specialized expertise, a duty ethic characterizing officials, and a predictable set of general rules.[4] Typically, bureaucrats are contractually obligated to the organization, work full time under its control, and identify their careers with it, although recent studies of U.S. public service suggest that these traditional descriptions are changing.[5] Nonetheless, the above characteristics of bureaucratic structures help us understand the nature of administrative agencies and their operations.

Federal administrative agencies derive their powers from Congress and the president and, ultimately, the people, who delegate tasks to them to be sure the business of government is carried out. Agencies are most often created through enabling statutes that delegate powers to them so they may advance the public interest. In addition to creating administrative agencies, enabling legislation often specifies the agency's location, resources, longevity, authority, and the means of exercising its delegated powers. The president has reorganization powers to shift, merge, or close down agencies, subject to certain limitations.[6] While administrative agencies may be referred to as a fourth branch of government, this is not accurate because they lack the independence of traditional branches, remaining under the control of the three constitutional entities. Nonetheless, it is estimated that "over 90 percent of the laws that regulate our lives, whether at work or at play, are now made by our public administrators, not by our legislators or our traditional lawmakers."[7]

In completing delegated tasks, bureaucrats, or public servants, exercise considerable administrative discretion. The need to control bureaucratic discretion to avoid threats to representative government was recognized by Alexander Hamilton and James Madison, among others.[8] These and other early Americans expressed concern that delegating too much power to appointed officials, who are not subject to electoral control, is a potential threat to democracy. In response, Congress decided to require that certain administrative decision-making processes remain open to public participation.

Additional constraints on agency and bureaucratic discretion include limits set by legislators and the courts; civil servants' political or ideological dedication to serving the interests of citizens; and professional administrative standards, which imbue public servants with the desire to behave ethically when making public decisions.[9] These various limitations, together with other bureaucratic characteristics (e.g., written documentation or "red tape"), help keep appointed bureaucrats accountable to the people, albeit indirectly and not always successfully.

Those working in administrative agencies are part of the "doing" side of government, that is, they implement the objectives of the organization. They translate general laws into more specific rules, regulations, and bureaucratic routines. In doing so they engage in two other important functions: rule making and administrative adjudication. Rule making, sometimes referred to as secondary legislation, involves establishing standards that can be applied to a class of individuals or an industry. Administrative adjudication, or "order making," involves the application of rules to specific individuals or firms to resolve disputes with regulated parties. Both functions are significant because in each case civil servants are often required to both interpret and implement the law, in which case they are essentially engaging in lawmaking.[10] Thus, administrative agencies have legislative powers (rule making), judicial powers (adjudication), and also executive powers (investigating misconduct). Consequently, they are crucial participants in the policymaking process.

Environmental Agencies and Public Policymaking

Several federal administrative agencies deal with the environment in one way or another, including thirteen of the fifteen cabinet-level departments.[11] Five agencies in particular (two of them cabinet level), are especially important in the institutional context of environmental policy implementation: the Environmental Protection Agency (EPA), the Department of Interior (DOI), the Department of Energy (DOE), the Nuclear Regulatory Commission (NRC), and the Council on Environmental Quality (CEQ). The EPA is the most important agency affecting environmental matters. The EPA works with its state counterpart agencies to implement and enforce environmental protection laws. If states fail to meet their responsibilities, the EPA will intervene.

Before briefly considering these agencies, some preliminary distinctions about agency types are in order. Agencies can be classified as executive, independent, or hybrid. Executive agencies are those headed by an administrator who is appointed by the president, with the advice and consent of the Senate. They are typically located in one of the fifteen cabinet-level departments (e.g., the Occupational Safety and Health Administration is within the Department of Labor). Independent agencies are not part of a cabinet department; they are headed by commissioners appointed by the president with the advice and consent of the Senate, who serve for fixed terms of office and can only be removed for cause (e.g., Nuclear Regulatory Commission). Hybrid agencies, sometimes referred to as independent executive agencies, are not located within an executive branch department (e.g., the Environmental Protection Agency, Federal Energy Regulatory Commission). A single administrator appointed by the president heads each agency and reports directly to the president rather than to a department-level secretary. Keeping these distinctions in mind, let's briefly introduce four important environmental agencies (the EPA will be explored in depth later in the chapter).

The Department of the Interior, a cabinet-level executive department, was created in 1845. Its responsibilities are to restore and maintain the health of public lands, natural resources, and waters under federal management so that they are used and developed in ways that are environmentally appropriate. It seeks to achieve a balance between natural resource preservation and economic growth—a formidable task given the competing interests of those who prefer expanded use of natural resources (e.g., timber, mining, recreation interests) and those seeking to preserve natural resources. It also has responsibilities for preservation of plant and animal species and habitats.

A cabinet secretary heads the DOI. Presidential appointments to this position have seldom encountered Senate opposition. Indeed, there have been only four instances where significant opposition occurred on confirmation votes for the Secretary of the Interior: votes on Walter Hickel (1969), Stanley Hathaway (1975), James Watt (1981), and William Clark (1983).[12]

The DOI's 2015 annual budget was $12 billion, with a staff of 72,204.[13] Included as subunits within the DOI, among others, are the Bureau of Land Management, the National Park Service, the Bureau of Reclamation, the Bureau of Indian Affairs, and the U.S. Fish and Wildlife Service, all vitally important to environmental protection efforts. The Bureau of Land Management administers more than 245 million acres of public lands, mostly in a dozen western states.[14] It works to sustain the health and diversity of these lands for public use and enjoyment. The National Park Service promotes and regulates the use of national parks. Its purpose is to conserve natural and historic objects, scenery, and wildlife within the parks. The Bureau of Reclamation manages, develops, and protects water and water-related resources. It seeks to protect local economies and preserve natural resources through effective use of water. The Bureau of Indian Affairs is concerned with promoting economic opportunity and protecting trust assets of American Indians and their tribes. The U.S. Fish and Wildlife Service is primarily responsible for fish, wildlife, and plant preservation, including migratory birds, endangered species, and marine mammals.

The Department of Energy was created as a cabinet-level executive department in 1977. In 2015 it had a budget of $27.9 billion with $22 billion in annual contract obligations and a staff of 15,832 federal employees.[15] DOE takes the lead in promoting diverse energy sources, efficient energy use, and improved environmental quality, among other things. It has an important role in promoting science and technology and in national security, but for our purposes its role in energy and environmental matters is the chief focus. DOE encourages energy efficiency by exploring new energy-related technologies, increasing customer choice of energy sources, and ensuring adequate and clean energy supplies. It also aims to minimize U.S. vulnerability to events that could reduce energy supplies. DOE seeks to improve environmental quality by controlling risks and threats (e.g., safety, health, environmental) from agency actions, cleaning up contaminated areas, and developing new technologies for ameliorating environmental problems.

Personal Profile

Steven Chu: The Scientist in Charge

Physicist Steven Chu, a Nobel laureate, served as U.S. Secretary of Energy under President Barack Obama from 2009 to 2013. Dr. Chu was the first person nominated to the U.S. cabinet who had been a Nobel Prize recipient, the first scientist to head DOE, and the second Chinese American to serve in a U.S. cabinet role.[1] He is also the longest serving energy secretary.[2] Chu won the Nobel Prize for Physics in 1997, for his work on cooling atoms with lasers.[3] Prior to joining the cabinet, he was a professor of physics and molecular and cellular biology at the University of California, Berkeley, and director of the Lawrence Berkeley National Laboratory. Before that he was a professor of physics at Stanford and following his service as energy secretary he returned to his faculty post at Stanford.[4]

As secretary of energy, Chu vigorously advocated research on renewable energy and nuclear power. He believed combating climate change required a shift away from fossil fuels.[5] Upon leaving the cabinet he warned of the threat posed by climate change: "If we don't change what we're doing, we're going to be fundamentally in really deep trouble. We're already in trouble. So we have to transition to better solutions."[6] He cautioned against continued reliance on fossil fuels, noting, "As the saying goes, the Stone Age did not end because we ran out of stones; we transitioned to better solutions."[7] In discussing the effects of climate change and strategies for addressing it, he has asserted that California farms could be wiped out by the end of the century because of global warming, and that a nuclear power plant emits one hundred times less radiation than a typical coal-burning power plant.[8]

While leading the DOE he was the force behind ARPA-E (Advanced Research Projects Agency-Energy), the Energy Innovation Hubs, and the Clean Energy Ministerial meetings.[9] He is credited with creating a constructive, collegial "Bell Labs" culture in ARPA-E that ultimately diffused to other parts of DOE such as the solar photovoltaic program, "SunShot."[10] ARPA-E sought to support high-risk, high-reward technology. During his four-year term, solar energy deployment increased tenfold, and renewable energy deployment doubled in the United States.[11] President Obama called upon Chu to help BP stop the Deepwater Horizon oil leak in the Gulf of Mexico. He was similarly asked to assist the Japanese government in responding to the damaged nuclear reactor following the tsunami at Fukushima-Daiichi and to assist FEMA in responding to Hurricane Sandy.[12] He supported Obama's agenda to invest in clean energy, seek solutions to the global climate crisis, cultivate partnerships with industry, increase energy efficiency, curb reliance on foreign oil, and help create new employment opportunities.

Dr. Chu's accomplishments have been widely recognized, with numerous honors. He is a member of the United States National Academy of Sciences, the American Academy of Arts and Sciences, the American Philosophical Society and the American Sinica, among other prestigious groups.[13] He is the recipient of twenty-three honorary degrees, including honorary doctorates from Harvard, Boston University, Washington University-St. Louis, Yale, New York University, Penn State University, and many others.[14] As a scientist, administrator, and vocal advocate, Dr. Chu has contributed greatly to our understanding of global environmental issues and taken steps to address these issues at the highest levels of government.

Notes

1. Sky Canaves, "Commerce Nominee a Locke in China," *The Wall Street Journal* (China Edition), February 26, 2009; http://blogs.wsj.com/china realtime/2009/02/26/commerce-nominee-a-locke-in-china/. Accessed November 20, 2014.

2. "Stephen Chu," Stanford University (Department of Physics Faculty); https://physics.stanford.edu/people/faculty/steven-chu. Accessed November 17, 2014.

3. "Stephen Chu: Government Official, Physicist, Scientist," The Biography. com website; http://www.biography.com/people/steven-chu-9247820. Accessed November 20, 2014.

4. Stanford University, "Stephen Chu."

5. H. Josef Hebert, "Energy Secretary Pick Argues for New Fuel Sources," *Seattle Times*, December 11, 2008; http://seattletimes.com/html/politics/2008492740_apchuprofile.html. Accessed November 20, 2014; Sarah Jane Tribble, "Nuclear: Dark Horse Energy Alternative," *Inside Bay Area: Oakland Breaking News*, June 18, 2007; http://www.insidebayarea.com/ci_6168586. Accessed November 20, 2014; Michael Anastasio, Samuel Aroson, Steven Chu, John Grossenbacher et al., "A Sustainable Energy Future: The Essential Role of Nuclear Energy," *Department of Energy*, August 2008; http://www.energy. gov/node/28891/pdfFiles/rpt_SustainableEnergy Future_Aug2008.pdf. Accessed November 20, 2014.

6. Bjorn Carey, "Q&A: Steven Chu on Returning to Stanford, his Time as U.S. Energy Secretary," *Stanford Report*, May 15, 2013; http://news.stanford.edu/news/2013/may/steven-chu-qanda-051513.html. Accessed November 17, 2014.

7. Steven Chu, "Letter from Secretary Steven Chu to Energy Department Employees," *Department of Energy*, February 1, 2013; http://energy.gov/articles/letter-secretary-steven-chu-energy-department-employees. Accessed November 20, 2014.

8. Keith Johnson, "Steven Chu: 'Coal is My Worst Nightmare," *Wall Street Journal*, December 11, 2008'http://blogs.wsj.com/environmentalcapital/2008/12/11/steven-chu-coal-is-my-worst-nightmare/. Accessed November 20, 2014; Joe Romm, "Steven Chu's Full Global Warming Interview: 'This is a Real Economic Disaster in the Making for our Children, for your Children," *Climate Progress*, February 9, 2009; http://think progress.org/romm/2009/02/09/203665/stephen-chu-la-times-interview-global-warming/. Accessed November 20, 2014.

9. Carey, "Q&A: Steven Chu."

10. Stanford University, "Stephen Chu."

11. Carey, "Q&A: Steven Chu."

12. Stanford University, "Stephen Chu"; Stephen Chu, "Secretary Steven Chu Announces Departure, Accomplishments in a Letter to Energy Department Employees," *Renewable Energy World.com*, February 1, 2013; http://www.renewableenergyworld.com/rea/news/article/2013/02/secretary-steven-chu-announces-departure-accomplishments-in-a-letter-to-energy-department-employees. Accessed November 17, 2014.

13. "MIT World Speakers: Steven Chu," Massachusetts Institute of Technology Compton Lectures, last modified August 2008; http://compton.mit.edu/speakers/steven-chu/. Accessed November 20, 2014; "Dr. Steven Chu—Former Secretary of Energy," Department of Energy (About Us); http://energy.gov/contributors/dr-steven-chu. Accessed November 17, 2014.

14. "Steven Chu," Wikipedia, last modified October 10, 2004; http://en.wikipedia.org/wiki/Steven_Chu. Accessed November 17, 2014.

The Nuclear Regulatory Commission, an independent agency, was created in 1974. Its budget in 2015 was $1,059.5 million.[16] The NRC's mission is to ensure the public health and safety and protection of the environment in the use of nuclear materials. Its broad responsibilities include regulating nuclear power reactors; transport, storage, and disposal of nuclear materials and waste; and the uses of nuclear materials (e.g., medical, academic, industrial). It issues licenses to construct and operate nuclear facilities and matters relating to nuclear materials (possessing, transporting, using, handling, and disposing of them). Five commissioners head the NRC, each appointed by the president and confirmed by the Senate, for five-year terms. One of them is appointed chairperson by the president.

The Council on Environmental Quality, an advisory unit to the president, was created by the National Environmental Policy Act (NEPA) of 1970. In 2015 its budget was $3,009,000.[17] The CEQ gathers and analyzes data, keeps the president informed about progress toward the goal of a cleaner environment, recommends environment-related legislation, and issues a public report annually on the state of environmental quality. It also aids federal agencies in meeting their responsibilities to complete environmental impact statements and environmental assessments.

This descriptive information about particular environmental agencies shows the differences that exist among such federal agencies, including structure, mission sources, and independence. Even more important is the role that administrative agencies play in the public policymaking process. We will examine four concepts or processes that help to explain their role: rule making, adjudication, iron triangles, and issue networks.

Rule Making

One way that administrative agencies make public policy, as stated previously, is to develop, change, and eliminate government rules. In doing so they follow procedures outlined in the Administrative Procedure Act (APA) of 1946. Congress, the president, and the judiciary have acted to ensure that rule making occurs in ways that are consistent with their priorities and values. The APA requires that agencies inform the public by a notice in the *Federal Register* of their intent to develop a rule, inviting public comment. Subsequently, agencies assemble the required data, including comments from the public and other interested parties, to formulate a proposed rule or regulation, and again publish a notice in the *Federal Register* inviting public comment. Draft rules are also submitted to the White House Office of Management and Budget (OMB) for its review and approval. This duty to inform serves two purposes: it notifies the public of the rationale, purpose, and implications of the proposed rule, thereby enabling the citizenry to better participate in the process, and it offers a way for Congress, the president, and the judiciary to hold agencies accountable.

Public participation involves the submission of written comments on proposed rules. The APA seeks to prevent preferential access to the rule-making process and allows for other opportunities for public involvement (e.g., public hearings, consensual rule development, cross-examination of agency rule makers). The exact nature of the participation depends on the formality of the rule-making process.[18]Accountability is maintained under APA by judicial review of both the substance and process of rule making, by congressional oversight of the rule-making process (e.g., budget, investigation, appointments), and by presidential oversight of the process (e.g., submissions of regulatory agendas, OMB review of proposed and final rules). Once the agency has received and considered public comments and other related materials submitted to it by interested parties, the final rule is published in the *Federal Register* together with the agency's response to important issues raised in connection with the review. (A somewhat similar process is found when states' agencies engage in rule making.)[19]

Case Study

EPA Regulation of Carbon Emissions

In June 2014, the Environmental Protection Agency proposed a plan for new regulations to cut carbon pollution from power plants. The EPA is legally empowered to regulate carbon dioxide under the 1970 Clean Air Act, according to a 2007 Supreme Court ruling in *Massachusetts v. EPA*.[1] The Court's 5–4 decision was a rebuke to the Bush administration, which asserted that EPA lacked the authority to regulate carbon dioxide under the statute, and an opening for President Obama to take executive action.[2] Politicians from coal states oppose such rules, but more than half of the public support them even if it means higher electricity bills.[3] Obama was unable to get the Senate to pass a cap-and-trade bill in his first term (see chapter 4), and congressional opposition then prompted him to see what he could achieve acting on his own using Court-approved executive power through the EPA. The president directed the EPA to issue the rule by mid-2015, thereby enabling him to begin enforcement prior to the end of his second term.

Republican leaders and industry attorneys have vowed to stop or delay the EPA power plant rules by using budget riders and litigation, but Obama promises to continue pushing for his second-term environmental agenda, including sharp reductions of greenhouse gas emissions from new and existing power plants. Together with his "historic agreement" with China aimed at combating climate change, Obama views the rule as essential to limit heat-trapping greenhouse gases linked to global warming.[4] He seeks to reverse two decades of relative inaction on climate change by the United States.[5] Working through the EPA, Obama's goal and that of the rule is a 30 percent reduction in carbon emissions from 2005 levels by 2030. This would affect the nation's six hundred coal-fired power plants, the nation's largest source of carbon pollution, by 2030. Opponents view the action as an imperial overreach and a "war on coal."

The conflict over the EPA regulations reflects the red state versus blue state divide. Several red states have a vested interest in protecting their manufacturing centers, many of which are linked to fossil fuel production and use, including oil, natural gas, and coal. By contrast, many blue states are less reliant on fossil fuel and coal as they have transitioned toward a postindustrial economy.[6] EPA regulators have to navigate in these treacherous political waters as they craft rules that grant state flexibility while simultaneously imposing reductions on both high-emitting red states and low-emitting blue states. States have flexibility in deciding how to meet the standards, such as mandating that power plants cut emissions (e.g., shift to a fuel source with lower emissions) or upgrade equipment or efficiency.[7] The two main components of the EPA rule include state-specific carbon emission–based goals and guidelines with timetables for developing, submitting, and implementing state plans.[8] Blue states are moving toward a lower-carbon economy and are more likely to follow the path of California's successful cap-and-trade program to reduce emissions (see chapter 2).[9]

The timing of the rule coincides with renewed attention to international efforts to address climate change, prompted by the release of the report from the American Association for the Advancement of Science that warns of the consequences of human-caused climate change (e.g., rising sea levels [see chapter 2], extreme droughts and heat waves, food and water shortages, and stronger storms) and the announcement of the U.S.-China agreement to take action on the issue.[10] The timing of the new rule is in sync with the plan set forth in a United Nations accord in 2009, when Mr. Obama pledged the United States would cut its greenhouse gas pollution 17 percent by 2020, and 83 percent by 2050.[11] It is also a forerunner to the 2015 United Nations Conference of the Parties on Climate Change in Paris, France, which many hope will result in a binding global warming treaty.

This reliance on individual states to develop and implement federal environmental policies is not universally applauded. George Gonzalez (2015) argues that placing responsibility on the states to act under this rule is largely symbolic politics because it often weakens or undermines policies, given the states' tendency to prioritize economic growth over environmental protection. Gonzalez maintains that states cannot be depended upon as reliable partners in pollution abatement because of their natural inclination to advance an economic growth agenda, and that ultimately they enable the federal government to evade regulatory responsibilities.[12] Other observers tout state flexibility under the EPA rule as "the glue that holds the plan together" and pivotal to success of the regulatory initiative, making it easier and more cost effective for states to comply by adopting polices best tailored to their economy and energy mixes.[13] Several federal entities including the EPA, U.S. Department of Energy, Federal Regulatory Commission, and U.S. Department of Agriculture, among others, will assist the states as they implement their plans.[14] Furthermore, if a state fails to come up with an effective implementation plan, a federal plan can be imposed by the EPA.[15]

Supporters and opponents of the rule assess the benefits and costs of compliance differently. EPA and environmental activists highlight likely savings of $48 billion to $90 billion by 2030 attributable to improved energy efficiency, plus health and environmental gains from curbing emissions, while business and industry groups represented by the U.S. Chamber of Commerce focus on estimated costs to the economy of up to $50 billion annually and elimination of 225,000 jobs.[16] With stakes this high—hundreds of possible plant closures, possible system changes to the American electric power industry, transformation of how power is generated and used, large sums of money saved or spent, powerful environmental impacts, and gains or losses in health and employment—the battle lines are drawn. Renewable energy producers (solar and wind sector) and some utilities will stand to benefit; the coal industry and its allies together with many business groups are adamantly opposed. Legal and legislative attacks are expected to be numerous and drawn out.[17]

Notes

1. Environmental Protection Agency, "Proposed Endangerment and Cause or Contribute Findings for Greenhouse Gases under the Clean Air Act," last modified September 9, 2013; http://www.epa.gov/climatechange/endangerment/archived. html. Accessed November 20, 2014; Linda Greenhouse, "Justices Say E.P.A. Has Power to Act on Harmful Gases," *New York Times*, April 3, 2007; http://www. nytimes.com/2007/04/03/washington /03scotus.html?pagewanted=all. Accessed November 20, 2014.

2. Darren Samuelsohn, "The Greening of Barack Obama: He Didn't Set Out to be an Environmental President. He is Now," *Politico Magazine*, November 18,

2014; http://www.politico.com/story/2014/11/barack-obama-environment-112974. html. Accessed November 20, 2014; Robert Barnes and Juliet Eilperin, "High Court Faults EPA Inaction on Emissions," *The Washington Post*, April 3, 2007; http://www. washingtonpost.com/wp-dyn/content/article/2007/04/02/AR2007040200487.html. Accessed November 20, 2014; Juliet Eilperin and Steven Mufson, "Everything You Need to Know About the EPA's Proposed Rule on Coal Plants," *The Washington Post*, June 2, 2014; http://www. washingtonpost.com/national/health-science/epa-will-propose-a-rule-to-cut-emissions-from-existing-coal-plants-by-up-to-30-percent/2014/06/02/f37f0a10-e81d-11e3-afc6-a1dd9407abcf_story.html. Accessed November 20, 2014.

3. Coral Davenport, "In Climate Deal with China, Obama May Set 2016 Theme," November 13, 2014; http://www.nytimes.com/2014/11/13/world/asia/in-climate-deal-with-china-obama-may-set-theme-for-2016.html. Accessed November 20, 2014.

4. Josh Lederman, "How Obama's Power Plant Emission Rules Will Work," Associated Press, May 31, 2014; http://www.huffingtonpost.com/2014/05/31/obama-power-plant-emission-rules_n_5423726.html. Accessed November 20, 2014.

5. Coral Davenport, "Governments Await Obama's Move on Carbon to Gauge U.S. Climate Efforts," *New York Times*, May 27, 2014, A-11.

6. Ronald Brownstein, "The Politics of Being Green: How EPA's Proposed Regulations on Carbon Emissions will Exacerbate the Geographic Red-Blue Divide," *National Journal*, June 5, 2014; http://www.nationaljournal.com/political-connections/the-politics-of-being-green-20140605. Accessed November 20, 2014.

7. Kate Sheppard, "EPA Releases Much-Anticipated Limits on Power Plant Emissions," *Huffington Post*, June 2, 2014; http://www.huffingtonpost. com/2014/06/02/epa-carbon-rules_n_5428632.html. Accessed November 20, 2014; Coral Davenport, "Obama to Take Action to Slash Coal Pollution," *The New York Times*, June 2, 2014, A-1.

8. Federal Register, "Carbon Pollution Emission Guidelines for Existing Stationary Sources: Electric Utility Generating Units," June 18, 2014; https://federalregister.gov/a/2014-13726. Accessed November 20, 2014.

9. Davenport, "In Climate Deal."

10. Davenport, "Governments Await Obama's Move."

11. Lederman, "How Obama's Power Plant."

12. George A. Gonzalez, "Is Obama's 2014 Greenhouse Gas Reduction Plan Symbolic?: The Creation of the U.S. EPA and a Reliance on the States," *Capitalism Nature Socialism*, Forthcoming (2015).

13. Sheppard, "EPA Releases Much-Anticipated Limits"; Davenport, "Obama to Take Action."

14. Federal Register, "Carbon Pollution Emmission."

15. Sheppard, "EPA Releases Much-Anticipated Limits"; Eilperin and Mufson, "Everything You Need."

16. Ibid.; Davenport, "Obama to Take Action."

17. Robert F. Durant, "Linking Problems, Policy, and Public Management," in *Why Public Service Matters: Public Managers, Public Policy, and Democracy* (New York: Palgrave Macmillian, 2014), 53; Davenport, "Obama to Take Action."

Adjudication

Another way administrative agencies make public policy, in addition to rule making, is administrative adjudication of specific cases. Adjudication is conducted under provisions of the APA. When it makes its judgment, a public agency determines the winner and loser and issues an order to one of the parties in the dispute. Orders resulting from administrative adjudication are directed to the disputants in the conflict, the general public is excluded from participation, and adjudicative facts, not legislative facts, are the bases for deliberation. Administrative orders are designed to address past disputes, not to give general policy guidance for the future; nonetheless, orders do set precedents. What occurs is a case-by-case approach to creating regulatory policies.

While APA legislation spells out the procedures to be followed for both rule making and adjudication, the requirements for each differ. Adjudication involves dispute resolution of past behavior and results in policies directed at specific named parties in a dispute. By contrast, rule making aims to control future conduct by regulating parties in general.[20] In reality, these conceptual distinctions are not as neat as they may appear, and often there is confusion about which form of discretionary authority administrative agencies should exercise.

Iron Triangles

Yet another way to analyze the role of administrative agencies in the policy process is to consider the avenues of access available to those wishing to influence the formulation and adoption of policies. Access refers to the actual inclusion of various interests in the decision-making process. Access and its influence in policymaking can be explained by considering the function of "iron triangles" and "issue networks." Iron triangles, or subgovernments, refers to the cooperative, stable relationships that exist among participants in the policy process, specifically an administrative agency, a congressional committee, and related interest groups. Years ago, Douglas Cater described iron triangles in this way:

> In one important area of policy after another, substantial efforts to exercise power are waged by alliances cutting across the [executive and congressional] branches of government and including key operatives from outside. In effect, they constitute subgovernments of Washington comprising the expert, the interested and the engaged.[21]

Each of these participants represents a point on the triangle, and together this triad helps shape policy in a particular domain. For example, in the nuclear energy policy domain during the 1950s and 1960s the iron triangle consisted of the Atomic Energy Commission (an independent agency in the executive branch), the Joint Committee on Atomic Energy (in Congress), and key interest groups from the nuclear power industry (e.g., General Electric, Westinghouse, Combustion Engineering, Babcock & Wilcox).[22] Together this triad of participants influenced the development of nuclear energy policy: an iron triangle might include administrative agencies (DOI, DOE, NRC), congressional committees (Senate Energy and Natural Resources, Environment and Public Works; House Commerce, Resources, and Science Committees), and particular interest groups (e.g., American Public Power Association, National Coal Council, Environmental Defense Fund, Sierra Club).[23]

Issue Networks

Issue networks are composed of those concerned with a particular policy area who share common interests or stakes in decisions. These include elected and appointed officials, consultants, policy experts, activists, and interest groups. Given the proliferation of interest groups, activists, and policy experts in recent years, especially surrounding highly technical matters such as nuclear energy, the term *issue networks* provides a better description of the current policy participants. The issue network around nuclear energy consists of the four key environmental agencies at the core—EPA, DOI, DOE, and NRC—with four standing committees in the Senate and House, and ten or eleven other, more loosely related groups having access, involvement, and influence in the policy process.

It should be noted that turf battles are not unique to Congress, and such conflicts are clearly present in the interactions among environmental agencies as well. These are partially a result of overlapping jurisdictions, competing agency interests, and disagreement about goals. One consequence of such conflicts is that integration of environmental management is often lacking.[24]

Having briefly considered the missions and key functions of four federal environmental agencies, the rule-making and adjudication processes, the role of iron triangles and issue networks, and the potential for conflict among agencies, our attention turns to the most important federal agency, the EPA. We begin by briefly describing the internal organization of the EPA, and follow with more extended analysis of the external and internal environment of the agency, using "stakeholder theory" as a framework.

The Environmental Protection Agency: A Brief Profile

The EPA was created by President Nixon's executive order in 1970. The agency brought together various preexisting forms of pollution control into one federal regulatory unit. It has a broad and ever-expanding mission to protect, safeguard, and improve public health and the natural environment in the areas of air and water quality, and land use. Solid waste, pesticides, radiation, and toxic substance control fall under the jurisdiction of the EPA as well. Former EPA head William Ruckelshaus drew an analogy between the EPA's mission and efforts to give someone an appendectomy while the person is running the hundred-yard dash.[25] New pollutants or responsibilities are continually added to EPA's tasks, creating a moving target and making it difficult for the agency to complete one task before taking on another.

The EPA promotes its mission by implementing and enforcing federal environmental laws, integrating its efforts with those of other governmental units, and using the best available scientific information to reduce environmental risks. However, its budget rarely keeps pace with its responsibilities, making implementation and enforcement problematic.[26] Further, its ability to do high-quality scientific work has been hampered by understaffing and a declining capacity to conduct long-range environmental research.[27] The EPA does seek to inform major stakeholders about environmental and health threats and to engage them in a partnership to prevent or reduce such threats.

The agency's administrator reports directly to the president. It has issue-specific programs in air and water, radiation, solid waste, emergency response, and pesticides and toxic substances. Separate units exist for research and development, environmental information, enforcement and compliance, policy, economics and innovation, administration and resources, and international activities, allowing for some integration of activities.

Some EPA functions are centralized in Washington, D.C., while others are decentralized. EPA staff in Washington is responsible for rule making, but much of the environmental policy implementation occurs in the ten regional offices, where two-thirds of the agency staff is employed, and in state, tribal, and local governments. EPA staff has increased substantially from its first full year of operation (in 1971, with seven thousand staff) to the 15,997 staff in 2014. Similar increases can be seen in its budget, from $3.3 billion in 1971 to $7.92 billion for 2015.[28]

Currently, the EPA is composed of technical experts committed to environmental protection. During some time periods the agency has been criticized for being reactive (responding to public concerns); during other periods it is criticized for being too proactive (generating new policy initiatives).[29] Over the years a variety of proposals to reform the agency have been put forward by the Scientific Advisory Board, the General Accounting Office, the National Academy of Sciences, the National Academy of Public Administration, the Partnership for Reinventing Government, the Office of the Inspector General, and others.[30]

Probably the best single source of information on the EPA's current goals and performance targets is found in the EPA Draft 5-Year Strategic Plan (required by the 1993 Government Performance and Results Act, completed April 10, 2014, FY 2014–18).[31] It summarizes the agency's top five strategic goals and shorter-term objectives as well as specific accomplishments the agency intends to achieve over the next several years. For example, goal number one is to Address Climate Change and Improve Air Quality. Climate change is a "wicked policy problem," as Robert Durant points out, in part because it is interrelated with other problems, thereby requiring "cross-disciplinary, cross-jurisdictional, interorganizational, and cross-sectoral policy and management approaches."[32] The plan also specifies the means and strategies EPA will employ to accomplish its goals, and the external factors that may affect its ability to achieve its objectives. In addition, it considers high-priority programs that cut across EPA strategic goals and the ways that the EPA measures and assesses its progress. The plan indicates that the EPA is advancing its core values (science, transparency, and rule of law) and using creative, flexible, cost-effective, and sustainable actions to protect and improve human health and the environment.

In the next section we shift attention from the internal purpose, structure, resources, and initiatives of the EPA to its relations with those in its external environment.

Environmental Activists, Business Managers, and the Public: Perceptions of EPA Performance

Three key groups that affect and are affected by the EPA are environmental activists, business regulatory officers, and the general public. The Pew Research Center surveyed these three key stakeholders to solicit their views of the agency's performance and their support for the agency's mission. Favorability ratings distinguished between the EPA specifically and the federal government in general. All three stakeholder groups—the general public, business regulatory officers, and environmental advocates—gave the EPA better marks than the government as a whole. However, there are sharp differences in how the three groups evaluated the overall performance of the EPA, especially between environmental advocates (68 percent excellent/ good) and regulatory officers (41 percent excellent/good). One caveat noted in the study is whether respondents were really able to distinguish between the performance of the EPA in meeting its responsibilities and the performance of subnational environmental agencies. All three constituent groups were critical of the EPA for working too slowly and making their rules and forms too complex. However, criticisms of the bureaucratic process do not extend to criticisms of the EPA's employees: majorities in each stakeholder group indicate that the agency's employees are courteous and professional.[33]

Disparities among the three respondent groups were evident regarding assessments of the EPA's technical capability and its policy decisions. It is not a surprise that environmental advocates were much more supportive than those from business. The views of the agency's fairness and honesty were low, with most general public and business officer respondents saying that the agency gives preferential treatment to some groups. Environmental advocates have divided opinions on this issue. However, majorities in all three groups show greater confidence in the EPA's handling of safety issues than in the agency's fairness and honesty. Predictably, those who give the EPA high performance marks also view it more positively than do those who think the agency performs poorly. Respondents are more critical about the means (cumbersome bureaucratic processes) than the ends (policy priorities) in question. Regarding mission support, it is not surprising that respondents differ: two-thirds of the general public and eight out of ten environmental advocates support the EPA's mission; only four in ten business officers agree. As expected, mission support is correlated with favorability assessments. The report notes, "Support for the purpose of the agency even affects attitudes of environmental officers at manufacturing firms, who express reservations about the EPA in general. More than 80% of environmental officers who agree that strict environmental laws are worth the cost view the EPA favorably; only one-quarter of those who disagree with this trade-off hold a similar view."[34]

It is clear that these three stakeholder groups are important to the EPA's success. They can affect the agency's operations, but their interests often differ—and they vary in their judgments about the agency's performance. Administrative agencies such as the EPA need to identify the interests of such stakeholders, to consider how and when to act toward them, and adjust their agency's priorities to take into account stakeholder preferences. Sometimes agencies succeed at this task, other times they fail. If stakeholders are supportive rather than antagonistic, it can greatly influence the ability of the agency to advance its initiatives. However, it is not always easy to respond to competing and sometimes contradictory claims and to tailor strategies in order to address the interests of multiple constituencies. This balancing act in satisfying diverse claimants is a test of the effectiveness of agencies that serve "multiple constituencies." Satisfying these diverse interests is especially complicated when attempting to respond to the conflicting interests and agendas found in the U.S. Congress. This complexity is related to the committee structure of Congress: in the 110th Congress (2009–11), seven full Senate committees, including thirteen subcommittees, and eight full House of Representatives committees with eighteen subcommittees have EPA jurisdiction.[35]

The public is an especially important stakeholder for government agencies like the EPA to consider because we expect policymakers to respond to fluctuations in public sentiments.[36] Survey data collected by different sources, on public sentiments, yield the following results:

- 57% have a very favorable or mostly favorable opinion of the EPA, 33% very unfavorable or mostly unfavorable;[37]

- 39% rate the job being done by the EPA as excellent or good, 21% poor;[38]

- 40% have very positive or somewhat positive feeling toward the EPA, 28% very negative or somewhat negative;[39]

- 69% favor the EPA's updating standards with stricter limits on air pollution, 26% strongly or somewhat oppose;[40]

- 72% believe the EPA's proposed standards to reduce carbon emissions from power plants will have a positive effect, 7% believe the impact will be negative. After listening to a debate for and against the new standards, 63% were in favor and 33 opposed;[41]

- 62% strongly trust or somewhat trust the EPA as a source of information about global warming, 38% strongly distrust or somewhat trust EPA information.[42]

We do not know the extent to which those surveyed were knowledgeable about each of the issues examined. It is clear from other surveys that most Americans do not fully understand the nuances involved in environmental policy and regulatory actions. Such high-tech issues serve to limit public participation in a democracy. As Edwards and his colleagues note, it is a continuing challenge during policy debates for government agencies such as the EPA to maintain a balance between technological expertise and participation.[43] Agencies like the EPA are a repository of scientific expertise, but agency employees, important internal stakeholders, must be mindful of the public's desire for input on even highly complex technological issues. The discussion now shifts to consider the views of EPA employees.

EPA Employees as Internal Stakeholders

EPA employees are another key stakeholder group crucial to the performance of the agency. The 2014 Federal Employee Viewpoint Survey (EVS) results provide data on the attitudes and perceptions of EPA and other federal employees in important areas of their work experience.[44] The satisfaction and commitment of agency staff can influence mission success. The overall downward trend of responses to EVS questions from 2013 to 2014 may be attributable to budget cuts and furloughs, which adversely affect morale and productivity of EPA staff. Table 6.1, shows both the strengths and challenges reported by EPA agency employees.

Among the top five agency strengths identified in response to EVS questions are the staff's willingness to put in extra effort to get a job done, their work importance awareness, and their commitment to examine ways to improve job performance. The EPA average on each of these items in 2013 and 2014 ranged from 85 percent to 96 percent and diverged little from the 2014 overall federal average. When asked whether their supervisor talks with them about their performance and how they would rate the overall quality of work done by their work unit, EPA employees' responses were similar in 2013 and 2014, ranging from 85 percent to 88 percent; the EPA average on these questions was higher than the 2014 government-wide average.

Of the top five agency challenges identified by EPA employees, one showed improvement from 2013 to 2014, and the EPA average on each of the items was higher for EPA employees than for the 2014 federal average. Nonetheless, the percentages were much lower on these EVS questions than those reported for items representing agency strengths. Among the challenges are whether pay raises were dependent upon how well employees performed their jobs, whether employees had sufficient resources to get their job done, and whether the work unit is able to recruit people with the right skills. On these questions responses ranged from 47 percent to 59 percent for EPA staff. Lower percentage responses were also found when employees were asked whether steps are taken to deal with poor performers

Table 6.1. Strengths and Challenges Reported by EPA Agency Employees
(Average percent)

Employee Viewpoint Survey Question	2013 EPA	2014 EPA	2014 Federal
Strengths			
When needed I am willing to put in the extra efforts to get a job done.	95	96	96
In the last six months, my supervisor has talked with me about my performance.	87	88	77
I am constantly looking for ways to do my job better.	88	88	90
The work that I do is important	86	85	90
How would you rate the overall quality of work done by your work unit?	87	85	82
Challenges			
Pay raises depend on how well employees perform their jobs.	56	59	54
I have sufficient resources to get my job done.	51	49	39
In my work unit, steps are taken to deal with a poor performer who cannot or will not improve.	46	48	45
My work unit is able to recruit people with the right skills.	47	48	33
How satisfied are you with your opportunity to get a better job in your organization?	41	44	39

Adapted from "Best Places to Work in the Federal Government Rankings," Partnership for Public Service 2014

who cannot or will not improve and whether employees are satisfied with their opportunity to get a better job in their organization. Percentages on these questions ranged from 41 percent to 48 percent for EPA employees.

There were some other notable decreases since 2013 in EVS responses of EPA staff. One large decrease was found in the organization's leaders' lack of maintaining high standards of honesty and integrity (−8 percent). There were also slight decreases between EPA's 2013 and 2014 EVS responses on physical conditions allowing employees to perform their jobs well (from 73 percent to 68 percent), senior leaders generating high levels of motivation and commitment in the workplace (36 percent to 31 percent), and supervisors working well with employees of different backgrounds (65 percent to 60 percent). On the HCAAF (Human Capital Assessment and Accountability Framework) survey, positive responses to questions in each category were averaged to create an index score. On all HCAAF indices the EPA scored lower in 2014 compared to previous years, except for Job Satisfaction in 2013, which remained the same. These scores are similar to the government-wide scores for the time span reflecting equal trends.

In other areas, EPA employees stack up well in comparison to government-wide results. When asked a series of questions that gauge employee engagement (e.g., leadership, opportunity to use skills), EPA's 2014 Employment Engagement Index was equal to the government-wide result (63 percent). Similarly, questions on employee job satisfaction (e.g., satisfaction with jobs, pay, and organization), the EPA's 2014 Global Satisfaction Index was on par with the government-wide index (61 percent versus 64 percent). However, when considering questions used to create Best Places to Work rankings, the responses of both EPA employees and for workers across the federal government showed a downward trend from 2012 to 2014 (see Table 6.2). On these measures, the EPA rankings did decrease more markedly (9 percent to 13 percent) than those for the government wide rankings (4 percent to 5 percent) from 2012 to 2013. Nonetheless, most EPA and government-wide employees would recommend their organization as a good place to work and remain satisfied with their job and with their organization.

Table 6.2. Federal Government and EPA Workplace Satisfaction (Average percent)

Employee Viewpoint Survey Question	2012		2013		2014	
	EPA	Federal	EPA	Federal	EPA	Federal
Considering everything, how satisfied are you with your job?	70	68	63	65	61	64
Considering everything, how satisfied are you with your organization?	65	59	55	56	52	55
I recommend my organization as a good place to work.	75	67	66	63	63	62

Adapted from "Best Places to Work in the Federal Government Rankings," Partnership for Public Service 2013, 2014

To summarize, EPA employees as internal stakeholders are reasonably satisfied and committed to doing a good job. The 2014 EVS results show the EPA remained steady after a drop in 2013 due to fiscal constraints. The EPA performs well compared to overall federal agency averages, especially in employee satisfaction with pay and obtaining feedback from their supervisor. While there is a downward trend in some areas for EPA as well as government-wide employees, successes are occurring and challenges are confronted. A few problem areas stand out, such as the ability to recruit people with the right skill set, and the lack of sufficient resources to accomplish the agency mission. However, overall EVS responses fluctuate somewhat, but generally track closely with EVS responses from employees in other government agencies.

It is clear from the above examples that external and internal stakeholders differ in terms of their perceptions, power, and policy preferences regarding the EPA's efforts to regulate the environment. The political climate is continually changing as partisan control of the executive and legislative branches of government shifts from one party to the other, and the views of the public, business managers, environmental activists, and internal staff are often at variance when it comes to agency capabilities and actions taken by the EPA. This creates a complex political environment for EPA administrators, who must try to understand and predict stakeholder activity and gauge the legitimacy, urgency, and power of various claimants or constituency groups on specific issues. The above discussion of stakeholders does not adequately consider the role of the courts, other administrative agencies, state and local officials, scientists, academics, pollution control professionals, political parties, the media, or other influencer groups. It does highlight the political challenge facing agency administrators, who must manage their stakeholder relationships in a way that achieves the purposes of the agency. This challenge is not unique to the EPA, but is also encountered by managers of the Interior and Energy departments, the Nuclear Regulatory Commission, and to a lesser extent, the Council on Environmental Quality.

Conclusion

This chapter has highlighted the role of administrative agencies, the sources and limits of their power, the processes by which they make decisions, and the institutional context in which they operate. While there are a number of other agencies that formulate, implement, and enforce environmental policies beyond those considered here, we have emphasized the important functions played by five key agencies, especially the Environmental Protection Agency. We indicated how administrative rule making and adjudication result in environmental policy, and how iron triangles and issue networks operate. Both the internal operating environment and selected external and internal stakeholders of the EPA were examined. We showed how Congress and the president can influence the actions of administrative agencies, and how the public, regulated business, environmental activists, and EPA staff view the agency and the issue of environmental protection.

The mini-case study of the EPA's proposal for new regulations to cut carbon pollution from power plants illustrated the important legal and political issues that surround environmental regulation. The issue involved action by the president and the U.S. Supreme Court taken in part in response to inaction by the Congress. The EPA is exercising its regulatory role in initiating the regulation, the president is exercising his chief executive role in supporting the agency's action, majorities in Congress are vowing to stop or delay the rule, and the Court is exercising its interpretive role in clarifying the authority of EPA to regulate carbon emissions. The case also shows the role that state governments will have in implementing the rule. Supporters of the EPA rule saw it as a way to address the climate change issue and the health and environmental benefits to be gained, while opponents viewed it as regulatory overreach and stressed the costs to the economy and loss of jobs that would likely result.

The Environmental Protection Agency, like other administrative units, is actively pursuing alternatives to the costly and cumbersome command-and-control regulatory system. While top-down command-and-control regulations continue to be necessary, the EPA is simultaneously experimenting with a variety of approaches. It is also working on both pollution control and pollution prevention; however, progress on the prevention side has been slower to materialize. As policymaking becomes more complex, and iron triangles give way to issue networks, the EPA has responded with more emphasis on partnerships and strategies for dealing with external stakeholders. In response to the 1993 Government Performance and Results Act, the EPA has developed its five-year strategic plan, which includes many of these initiatives. Given the political climate in Congress, it is unlikely that major new environmental initiatives are going to be approved legislatively in the near term. Administrative agencies have the potential to revitalize environmental protection efforts, but to do so, agencies such as the EPA will need to continue to be responsive to current trends, to work with other agencies toward a system of integrated ecosystem management, and to reinvigorate their efforts through aggressive, protective actions.

Websites

Branches of Government

http://www.house.gov/content/learn/branches_of_government/

Agency List

https://www.federalregister.gov/agencies

Administrative and Regulatory Agencies that concern with the Environment

ENVIRONMENT PROTECTION AGENCY

https://www.epa.gov

FEDERAL ENERGY REGULATORY COMMISSION

https://www.ferc.gov

NUCLEAR REGULATORY COMMISSION

https://www.nrc.gov

OCCUPATIONAL SAFETY AND HEALTH ADMINISTRATION

https://www.osha.gov

DEPARTMENT OF INTERIOR

https://www.doi.gov

DEPARTMENT OF ENERGY

https://www.energy.gov

COUNCIL ON ENVIRONMENTAL QUALITY

https://www.whitehouse.gov/administration/eop/ceq

Subunits within the Department of Interior

BUREAU OF LAND MANAGEMENT

https://www.blm.gov

THE NATIONAL PARK SERVICE

https://www.nps.gov

THE BUREAU OF RECLAMATION

https://www.usbr.gov

THE BUREAU OF INDIAN AFFAIRS

https://www.indianaffairs.gov

THE U.S. FISH AND WILDLIFE SERVICE

https://www.fws.gov

A Guide to the Rulemaking Process

https://www.federalregister.gov/uploads/2011/01/the_rulemaking_process.pdf

Steven Chu

http://www.energy.gov/contributors/dr-steven-chu

EPA Regulation of Carbon Emissions

http://www.epa.gov/ghgreporting/

http://www.americaspower.org/sites/default/files/EPA%20Regulations%20
January%202015.pdf

Adjudication

ENVIRONMENTAL APPEALS BOARD

http://yosemite.epa.gov/oa/EAB_Web_Docket.nsf

Iron Triangles

http://examples.yourdictionary.com/iron-triangle-examples.html

Issue Networks

ENERGY AND WATER USE—CORPORATE ISSUE NETWORK

http://www.uschamberfoundation.org/initiative/energy-and-water-use

The Environmental Court

Introduction: The Peculiar Nature of the Court

The Supreme Court is clearly the most powerful court in the federal system, and possesses authority to hear and process judicial cases. At the same time, the Supreme Court is distinct in that not only does it interpret statutes and choose the cases it hears, it is also a generator of government policy. The Court is a political institution similar to other federal institutions, but unmistakably distinct from the Congress and the presidency in a number of ways.

So Why Is the Court Involved with Environmental Issues?

To enhance our understanding of the Court's treatment of environmental issues, we need to ask what factors influenced the Court to become involved with environmental issues in the first place. On the one hand, this involvement seems unreasonable since Supreme Court justices, on the basis of their own legal training and the experience of their staff, would not seem to be prepared to handle the complex and sometimes perplexing questions that the environment presents. Staff hired by each justice are few in number and similar in background, especially when compared to the average number of staff persons assisting individual House and Senate members in Congress. As of the October 2013 term, for example, each justice had no more than four law clerks, all selected from leading law schools across the country,[1] whereas House members could employ as many as eighteen persons of varying backgrounds,[2] and an individual Senator could employ as many as forty-five persons.[3]

The limited number of judicial staff persons and the uniformity of their training would seem to put some distance between the Court personnel and environmental concerns. This has been an apparent problem since the early 1900s, when the Court first reluctantly agreed to render a multistate judgment on pollution, in a case that illustrated the complexity of environmental issues. The justices found these decisions not to their liking.[4] They found it difficult to adapt standards they had used to cover legal inquiry to environmental questions, suggesting that judges are "unsuited to make policy decisions in technical areas such as pollution control because they must respond to individual demands for justice."[5] Moreover, the environmental movement has become increasingly more complex over time. Beginning early in the twentieth century with a concern for conserving resources, the environmental movement has branched out into much broader questions. Franklin Roosevelt's interest in the 1930s and 1940s, for example, was in preserving forests, purifying water and air, managing land, preserving wildlife, and creating and maintaining national parks and monuments. With the beginning of the more expansive environmental movement in the 1960s, after the publication of Rachel Carson's book *Silent Spring* in 1962,[6] questions regarding the risks to public health of uncontrolled use of pesticides were raised. From that time on, policymakers began looking at a more extensive range of issues such as those that concerned Bill Clinton in the 1990s, which included not only the particular conservation issues that concerned FDR, but also such issues as trade, wilderness areas, toxic waste, recycling, and environmental cleanup.[7]

The elaborate nature of the environment and its diversity poses a major problem for judges, as well as other policymakers involved with environmental questions. Nonetheless, despite these apparent barriers, there have been several reasons for the Court's involvement with environmental questions. One astute observer noted years ago that "[t]here is hardly a political question in the United States which does not sooner or later turn into a judicial one."[8] Alexis de Tocqueville, who made this perceptive observation, would have certainly agreed that environmental issues fit this category. This also suggests the importance of the courts in American government and politics, as they have been the forum for most of the important political questions we wrestle with today.

The Evolution of Judicial Involvement in Environmental Issues: Top-Tier Cases

Attention paid by the Court to the complex environmental issues is relatively recent. It has only been since the 1970s that environmentalists have made the Court their focus. Lettie Wenner argues that this was at the same time that "considerable skepticism had grown up around the ability of administrative agencies to carry out the goals of Congress." Thus, the Courts were used, Wenner maintains, "to force the Environmental Protection Agency and other governmental organizations to carry out the policies articulated in the laws."[9]

In the 1930s and 1940s environmental and conservation questions brought to the Court were infrequent and limited to fewer categories than one sees today. Furthermore, remedies relied on by the Court in resolving conservation questions were those with which they were already familiar. Such remedies included framing requests for environmental resolution in terms of trespass, personal injury, and liability for damages.[10] By the 1970s, we begin to see the Court more frequently approached to resolve environmental concerns, which caused Kenneth Holland to characterize this change as a period of *"judicialization"* of environmental policymaking.[11]

In order to assess and appreciate the complete involvement of the Supreme Court in environmental politics, we have divided our discussion of decisions the Supreme Court have rendered into two levels, namely, examining the most important top-tier case decisions and those we would call second-tier case decisions. Of the 150 Supreme Court cases decided for and against conservation and the environment since 1935, there have been a number of important top-tier cases that invite our close attention. These cases are considered top-tier because they have had the greatest impact on the political system, either by generating new federal regulations for courts and other federal institutions to follow, or in showing us the future direction the Court might take as a result of its decision regarding environmental matters. These top-tier decisions have also regularly redefined agencies and governmental bureaus in their roles as primary responders to environmental concerns. They have also, at times, settled disputes regarding the balance of power and authority among federal, state, and local jurisdictions when government has been involved with the environment. Often the top-tier cases have also involved significant and visible administration officials including the president, the vice president, and the secretary of the interior, as well as important environmental advisors to the president.

More specifically, the Supreme Court has found itself in a position to settle questions of conflicting power between the president and Congress when it has come to conservation and the environment. In *Train v. City of New York* (1975),[12] for instance, the president ordered the impoundment of environmental protection funds that had been earmarked for a program he vetoed. The Supreme Court sided with Congress, ruling that the president could not impound funds set aside for environmental protection since it would frustrate the will of the Congress by undercutting its program. In *Sporhase v. Nebraska ex re Douglas* (1982),[13] the Court dealt with a dispute over water being pumped between two states, indicating that water concerns were subject to congressional jurisdiction and regulation and not under the jurisdiction of states.

Since the 1970s, the Supreme Court has been consistent in its decisions recognizing the Environmental Protection Agency (EPA) Administrator and the agency as the primary responder to conservation and environmental concerns. This has not been without imposing some limitations on the agency, however. There have been fourteen cases since 1935 that were so decided. Included among these cases we find the *E. I. DuPont de Nemours & Co. v. Train* (1976)[14] case, where several chemical plants that released pollutants into the water disputed the EPA's authority to create industry-wide pollution regulations. The Supreme Court ruled that the EPA had that authority. In the 1992 *Arkansas v. Oklahoma*[15] case, the EPA granted Arkansas a discharge water permit. Oklahoma protested the particulars of the permit, but the Supreme Court indicated that if Oklahoma was dissatisfied with the standards specified in the permit, it would have to appeal to the EPA for any changes rather than the Court.

Alaska Department of Environmental Conservation v. EPA (2004)[16] saw the Supreme Court deciding whether the Alaska Environmental Agency's declaration that a company could use the "best available control technology" to conserve resources was valid or not, even though the technology approved was inferior to alternatives. The Supreme Court ruled that the EPA had authority to overrule any state agency's decision if, in its judgment, it did not see the state agency's decision as a valid process of applying technical knowledge.

At the same time, the Supreme Court has reserved to itself the right to determine standards for resolving pollution problems rather than looking to any other institution or agency. It has also retained the right to control the extent of environmental efforts as well as to define what an "emission" is. Over the years, ten decisions in particular have reiterated this power of the Court. Among the most important of them, we would call attention to the following: In the 1975 case of *Train v. Campaign Clean Water*,[17] for example, the Court showed that even though it has been protective of the EPA in establishing it as an instrument to carry out environmental laws, it would, on occasion, overrule the agency, as it did in this case when the administrator of the EPA refused to extend sufficient federal funds to states for environmental projects, as the Court felt it should. The Supreme Court also showed, in the case of *Adamo Wrecking Co. v. U.S.*(1978),[18] that it could take it upon itself to define what constituted an emission to be regulated, when it ruled against the government's allegation that the Adamo Wrecking Company, in demolishing a building, had created an emission (as defined by the EPA). And in 2007, the Supreme Court handed down a very important decision expanding the EPA's jurisdiction by ruling in *Massachusetts v. EPA*[19] that greenhouse gases were pollutants that were subject to regulation by the EPA, a finding that was reinforced by the Supreme Court's 2011 decision in *American Electric Power v. Connecticut*, [20] where the Court made it clear that only the EPA had the power to regulate greenhouse gases, inasmuch as it now controlled standards for emissions. In the April 2014 *EPA v. EME Homer City Generation, L.P. et al.*[21] case, the Supreme Court supported the EPA's authority to control emissions by coal-burning factories in twenty-eight Southern, Midwestern, and Appalachian states that pollute the air of states on the East Coast, although, in June 2014, in the case of *Utility Air Regulatory Group v. EPA et al.*, [22] the Court, while supporting the EPA, strongly criticized it for overreaching its authority. Justice Scalia's majority decision acknowledged that the EPA can limit greenhouse emissions from large industrial sources such as oil refineries and power plants, but denied the EPA permission to restrict emissions from smaller institutions such as small businesses, schools, and apartment buildings, suggesting that the agency exceeded its authority in controlling these emissions. The Court also warned the EPA not to overextend its authority beyond what the Clean Air Act allows in its language. As Scalia stated in some colorful language: "We are not willing to stand on the dock and wave goodbye as EPA embarks on this multiyear voyage of discovery . . . an agency may not rewrite clear statutory terms to suit its own sense of how the statute should operate."[23]

The Supreme Court has often shielded federal agencies, executive departments, and bureaus from having to comply with environmental restrictions. The following examples illustrate this point. Three cases involving the U.S. Navy allowed the Navy to avoid complying with environmental restrictions, including *Weinberger v. Catholic Actions of Hawaii Peace* (1981),[24] which indicated that the Navy could sidestep filing an Environmental Impact Statement, which would delay the storage of weapons and ammunition. In *Weinberger v. Romero-Barcelo* (1982),[25] the Navy was allowed to continue to discharge ordinance in the ocean despite the potential pollution the discharge would cause; and in the 2008 case of *Winter v. NRCD, Inc.*,[26] the Court allowed the Navy to continue to use sonar in training its sailors despite its serious threat to aquatic mammals.

The Supreme Court also allowed the secretary of commerce a free hand in deciding what should be done regarding Japan's refusal to comply with whaling restrictions, in the case of *Japan Whaling Assn. v. American Cetacean Society* (1986).[27] In addition, the Bureau of Land Management was protected by the Court from a lawsuit filed against it by the National Wildlife Federation concerning land use, in *Lujan v. National Wildlife Federation* (1990).[28] A final example involved the Department of the Interior, which received the Court's approval to sell oil and gas leases on the outer Continental Shelf off California, in the case of *Secretary of Interior v. California* (1984).[29]

In addition, the Supreme Court has resolved a number of interstate disagreements regarding the environment and conservation. Three such cases merit our attention. In *International Paper v. Ouelette* (1987),[30] officials at an International Paper Company mill located in New York State regularly discharged pollutants into Lake Champlain, which lies between New York and Vermont, with adverse effect on landowners on the Vermont side. The Supreme Court ruled that, as the source of the pollution was actually in New York, the Vermont state law under which the suit was filed did not apply. The second case involved a confrontation between two eastern states in the 2008 case of *New Jersey v. Delaware*.[31] Administrators at BP Oil Company wanted to construct a liquefied natural gas pipeline in New Jersey, which would require dredging underwater land in Delaware. Delaware protested, and the Supreme Court ruled that Delaware had the authority to deny such a permit. A final example occurred in 2011 in *Montana v. Wyoming and North Dakota*,[32] concerning the three states that shared water use. Montana sued Wyoming since it felt that Wyoming, in changing its system of irrigation, had reduced the wastewater returned to the Yellowstone River, depriving Montana's downstream appropriators of the water which they were entitled. Montana presumed Wyoming was using extra water beyond the agreement that had been made among the states. The Court ruled that it was not more water Wyoming was using; rather, the state was using a more efficient irrigation system to irrigate the same amount of water on the same size land mass.

The Supreme Court has also made a point of protecting the safety and public health of inhabitants of cities and states as well as protecting the environment in its decisions. Since the 1970s there have been ten cases that illustrate this. Of those, the most important include the case of *U.S. v. Penn. Industrial Chemical Corporation* (1973),[33] where the Court ruled that even though a company discharged pollutants that were not covered by a specific law, all pollution, whether mentioned in the law or not, is illegal. There were two additional cases protecting wetlands: *U.S. v. Riverside Bayview*

Homes (1985)[34] and *Border Ranch Partnership v. U.S. Army Corps of Engineers* (2002).[35] In both of these cases wetlands were safeguarded from potential damage as the Court ruled that government does have the authority to protect wetlands. In 1991 the Court ruled in the case of *Wisconsin Public Intervenor v. Mortier*[36] that a Wisconsin law that protected the public from pesticides was not preempted by the federal Insecticide, Fungicide and Rodenticide Act since the state law served as a more secure protection for its residents.

The Supreme Court has, moreover, been consistent in settling conflicts involving federalism, that is, in deciding whether federal, state, or local laws should prevail. Some nineteen cases were so decided between the years 1935 and 2013. The most important of these included five cases that ruled that federal law prevailed over conflicting authority, for instance, in *U.S. v. Oregon* (1935),[37] in which the state contested the use of land. The Court ruled that the federal government prevailed, since the land had previously been used as a federal bird sanctuary.

Native American rights have, at times, been in conflict with states' jurisdiction, since tribal reservations often overlap state lines. Two cases decided by the Supreme Court showed when the state might prevail over the rights of Native Americans. In the 1942 case of *Yulee v. State of Washington* (1942),[38] treaty power indemnified an individual Native American who was fishing without a state license from arrest, but in a 1968 state law, state regulation prevailed over the fishing rights of a member of the Puyallup Tribe in the case of *Puyallup Tribe v. Department of Game of Washington* (1968),[39] with the Court indicating that the state could limit fishing for all persons, even Native Americans.

Other protections available to the public were illustrated in the 1987 case involving coal mining limitations, *Keystone Bituminous Coal v. De Benedictis*.[40] Here the high court indicated that the Bituminous Mine Subsidence and Land Conservation Act, which limited coal mining, served to protect the public and must not be altered. In a 2005 case involving Alaska's attempt to protect wildlife on its land, state authority lost out to the federal government that wanted the land for its mineral leasing. In this case, *Alaska v. United States*,[41] the Court indicated that the land actually belonged to the federal government and not to the state. Finally, the Supreme Court declared that the Board of Harbor Commissioners of Los Angeles, in the case of *American Trucking Associations, Inc. v. City of Los Angeles, California* (2013),[42] could not demand that trucks comply with the board's Clean Air Action Plan, as the plan was preempted by the Federal Aviation Administration Authorization Act.

Other decisions uphold the power of state laws when they are challenged by companies. An example of this occurred in the 1974 case of *Air Pollution Variance Board v. Western Alfalfa Corporation*,[43] which originated when a state inspector measured the air quality of the alfalfa company without its knowledge or consent, from a location outside of company property. The Supreme Court ruled that even though the inspection occurred outside the company, it was a legitimate test to measure air quality. Another example that addressed state involvement consisted of one state discriminating against the solid and liquid waste from another state, in the case of *City of Philadelphia v. New Jersey* (1978).[44] The Supreme Court, in its decision, struck down the New Jersey law that would prohibit the importation of solid or liquid waste from outside the state, setting the precedent that no state could discriminate against another state's waste. A final example, from 1992, involved the case of *New York v. U.S.*[45] In this case the Court ruled that the Low-Level Radioactive Waste Policy Amendments Act, which tried to impose liability for waste generated within the borders of New York rather than merely attempting to encourage compliance, was unconstitutional.

The Evolution of Judicial Involvement in Environmental Issues: Second-Tier Cases

There have been several examples of second-tier case decisions rendered by the Supreme Court as well, but their consequences are limited compared to the first-tier cases. Nonetheless, it is important to take note of these decisions to understand the total involvement of the Supreme Court in environmental concerns. These areas of involvement have included three cases restricting certain activities in national parks and national forests. For example, in 1972, in *Sierra Club v. Morton*,[46] the Court ruled that it would not allow a portion of a national forest to be converted into a ski resort, thus preserving the protection government gives to national forests.

Six cases detail the proper reports, permits, and procedures that must be followed before an environmental case can be brought to the Court. For example, in the case of *Union Electric Co. v. EPA* (1976),[47] the Court ruled that states were free to enact stricter controls over companies violating the environment than national standards would require if the state saw this as necessary. In addition to these six cases, thirteen decisions have specified cleanup and payment procedures for damages to the environment. Three of these have involved oil spills, namely, *Askew v. American Waterways Operators, Inc.*(1973),[48] *U.S. v. Ward* (1980),[49] and *Midlantic National Bank v. New Jersey Department of Environmental Protection* (1986),[50] in which the Court in each case ruled that those responsible for the oil spill were required to pay the costs of the cleanup.

In addition to these cases, two decisions stipulated which parties in a conflict controlled private land, water, and/or minerals. In one of these cases, the *Andrus v. Shell Oil Company* case of 1980,[51] the Court had allowed citizens to purchase land that contained "valuable mineral deposits." When some people wished to buy land containing oil reserves and sought similar permission, they met with resistance, since oil had not initially been included among the "valuable mineral deposits." The Court settled this argument by listing "oil shale" as one of those mineral deposits, expanding the options to purchase land for citizens.

Another category of second-tier cases involved three Court decisions which included money appropriated to settle environmental matters, for example, the *Pennsylvania v. Union Gas Company* case in 1989.[52] The Union Gas Company appeared to be responsible for having created a hazardous waste site, and the federal government sued the company for money to clean up the area. The company complained, indicating that since the state of Pennsylvania owned the land, it should share in the cost of its cleanup. The Court agreed that that the state was liable to pay its share.

A final category of second-tier cases involves five decisions where the Supreme Court ended up protecting individuals as they confronted government. An example of this type of case is found in the 1979 case of *Kaiser Aetna v. U.S.*,[53] where the owners of a water pond turned it into a bay. The federal government wanted the bay to be a public bay, not a private one. The owners, however, went to Court suggesting that if the bay were made public it would be as if private property had been taken from the owners without compensation. The Supreme Court supported the owners' argument.

These second-tier cases all seem to have been limited to the specific individuals, companies, or situations they represented, without the broader consequences and ramifications of the first-tier cases. Moreover, they lacked the impact that the first-tier cases had on the political system. However, they do give an idea of how often the Court has been approached to render judgment concerning environmental questions.

Outcomes of Court Involvement in
Broadly Expanded Environmental Issues

While air, water, and hazardous waste cases still make it onto the agenda of the Supreme Court on a regular basis, it is the more diversified and mul- tifaceted environmental issues such as global warming and climate change, overpopulation, and ozone depletion that pose greater challenges for the Court. A number of these more complex environmental cases, decided since the 1970s, have divided the justices in various ways. Many were decided by close votes among Court members, including those dealing with lands, noise, emissions control, property, whaling, trade, and timber. Still other cases, mentioned previously, were decided by one-sided votes. Robert V. Percival, in his examination of environmental cases decided between 1970 and 1991, indicated that "more than a third of the environmental cas- es . . . were decided unanimously; more than 30 percent generated only one or two dissents while slightly less than one-third were decided over three or four dissents."[54]

It would appear, based on Percival's assessment of case decisions between the 1970s and 1990s, that the Court has become more supportive of the environment than opposed to it. So, does this mean that the Supreme Court is a reliable, eco-friendly institution that environmentalists and others can rely on in the future? Do those actual decisions supporting the environment suggest a firm commitment of the Court to the environment? We would submit that the answer is both yes and no. If we look at the current Supreme Court headed by Chief Justice John Roberts, for example, there are both those supporting the environment such as Associate Justices Ruth Bader Ginsburg and Stephen Breyer, and those opposed to the environment such as Associate Justices Antonin Scalia and Anthony Kennedy. For one legal scholar, William Funk, it was Associate Justice Stephen Breyer who attracted attention for his support of the environment. As he stated in his article entitled "Justice Breyer and Environmental Law," Breyer's record thus far would indicate that he is committed to "effective environmental law enforcement."[55]

Of those justices least supportive of the environment, Percival listed Antonin Scalia first, supporting the environment only 13.3 percent of the time in the fifteen cases he looked at, followed by Anthony Kennedy, supported the environment 28.6 percent of the time in seven cases.[56]

Scalia has made it quite clear, as he did in *Lujan v. National Wildlife Federation* in 1990, that he believes that the courts are not places to achieve environmental goals. Instead, people should look to other branches of gov- ernment to satisfy their need for reform.[57]

Personal Profile

The Environment: Friend and Supporter versus a "Slash-and-Burn" Opponent

The Roberts Court could well be labeled an anti-environmental court, given that there are typically five votes against environmental concerns, representing the Court's five most conservative members. That is why justices such as Ruth Bader Ginsburg are a breath of fresh air for environmentalists. Born in Brooklyn, New York, in 1933, Ginsburg quickly became a women's rights advocate and a defender of minorities as she was faced with blatant discrimination both in her education and in the workplace. Upon receiving her law degree from Columbia Law School in 1959, Ginsburg was employed as a law clerk, but by 1980 she had proven herself vital to the creation of the Women's Rights Project of the ACLU, and served as the American Civil Liberties Union's general counsel and on its national board of directors. She also won five of seven equal protection cases she brought before the U.S. Supreme Court. As a result of her views and her successes, she was nominated by President Bill Clinton to be an Associate Justice of the Supreme Court in 1993. Her actions to protect minorities caused President Clinton to describe her as a person who "repeatedly stood for the individual, the person less well-off, the outsider in society, and has given those people greater hope by telling them that they have a place in our legal system."[1]

Since Justice Ginsburg's elevation to the Court, she has moved to include environmentalists in her list of protected minorities. Perhaps her most noted contribution to such protections was her written opinion for the majority in the Supreme Court case of *Friends of the Earth v. Laidlaw Environmental Services* (2000).[2] When a group of citizens brought suit against a company for pollution, Ginsburg upheld the right of Friends of the Earth to sue, allowing more citizens to take action against industrial polluters.

Associate Justice Antonin Scalia, like most of his conservative counterparts, has a history of voting against environmental interests. However, Scalia is particularly opposed to environmentalists, as could be seen in his written opinion in *Lujan v. National Wildlife Federation* (1990),[3] which negatively affected citizens' ability to bring suit against anti-environmental agencies and companies that have damaged the environment.

Scalia, born in New Jersey in 1936, worked as a corporate lawyer for five years after graduating from Harvard University. He worked for the federal government from 1971 to 1977 as the general counsel of the Office of Telecommunications Policy (1971–72), chairman of the Administrative Conference of the United States (1972–74), and assistant attorney general for the Office of Legal Counsel (1974–77), serving in the Nixon and Ford administrations. In 1986, Scalia was nominated as an Associate Justice of the Supreme Court by President Ronald Reagan.

Though Justice Scalia typically casts his vote against environmental protections, one specific vote—and his written opinion for the majority—resulted in Justice Harry Blackmun stating in his dissent in *Lujan v. Defenders of Wildlife* (1992) that "I cannot join the Court on what amounts to a slash-and-burn expedition through the law of environmental standing."[4] He went on to suggest that "[i]n my view, the very essence of civil liberty certainly consists of the right of every individual to claim the protection of the laws, whenever he receives an injury."[5]

It wasn't until the year 2000 that Justice Ginsburg, in the previously mentioned *Friends of the Earth* case,[6] was able to change Justice Scalia's precedent with her own written opinion.

Notes

1. New York Times, "The Supreme Court: Transcript of President's Announcement and Judge Ginsburg's Remarks," June 15, 1993; http:www. nytimes.com/1993/06/15/us/supreme-court-transcript-president-s-announce-ment-judge-ginsburg-s-remarks.html?pagewanted=all&src=pm.
2. 528 U.S. 167 (2000).
3. 497 U.S. 871 (1990).
4. 504 U.S. 555, 606 (1992).
5. Blackmun here quotes *Marbury v. Madison*, 1 Cranch 137, 163 (1803).
6. 528 U.S. 167 (2000).

Given that many of the justices least sympathetic to the environment are in the majority on the Roberts Court today, it is hard to see the current Court as very friendly to the environment. This is one reason why the Court has been much more restrained in preserving the environment than even the lower federal courts.[58] Does this mean the Supreme Court will never be an eco-friendly institution? The answer really depends on those who sit on the Court. While the Court has never been an institution to lead on environmental decisions, especially when compared with Congress and the president, when the Court has been staffed with environmentally friendly justices, it has worked to broaden rules for standing, to make it easier for interested parties to bring environmental questions to the Court. Justices supportive of the environment have also made decisions that have had consequences in shaping the political system, as we pointed out in our listing of top-tier cases. In addition, the Supreme Court and other federal courts have played an important role in enforcing and defending environmental laws that have been the result of consequential legislation such as the National Environmental Policy Act (NEPA) of 1969. NEPA, one of the primary pieces of legislation defending the environment, specifies that "it is the continuing responsibility of the Federal Government . . . to improve and coordinate federal plans, functions, programs and resources," so as to "fulfill the responsibilities of each generation as trustee of the environment for succeeding generations" and to "attain the widest range of beneficial uses of the Environment without degradation, risk to health or safety, or other undesirable and unintended consequences."[59]

Since NEPA was introduced in 1969, and signed into law in 1970, there have been a number of other important environmental statutes designed to protect the environment that have been supported through Court action. William E. Kovacic, in his examination of the most important environmental legislation, lists a number of categories and specific legislation that the Court has supported, including laws designed to ensure that Americans can breathe clean air, such as the Clean Air Act and its amendments of 1990.[60] In addition, legislation has provided citizens with access to clean water, under the provisions of the Water Quality Act of 1987;[61] has limited ocean dumping, under the Marine Protection, Research and Sanctuaries Act of 1972;[62] has protected individuals from exposure to excessive noise in public places, in the Noise Control Act of 1972;[63] has safeguarded animals, as guaranteed in the Endangered Species Act of 1973;[64] has managed pesticides, through the Federal Pesticide Control Act of 1972;[65] has maintained land use in coastal areas as indicated in the Coastal Zone Management Act of 1972;[66] and has protected people from exposure to toxic waste, in the Toxic Substances Control Act of 1976.[67] In each of these instances, opinions filed by Supreme Court justices have legitimated the statute, solidifying protection of the environment as critical to society as a whole and advancing the environmental movement. Whenever the courts have opposed environmental protection in favor of development, as they have on occasion, environmentalism has been slowed in its expansion.

Regardless of the period under study, and which side the Court tends to support, its decisions are written without consensus or compromise—because it is a court. This, of course, potentially increases the likelihood of controversy in the community more than if the legislature alone should make a decision. In every Court decision some actors always win and some consistently lose, further politicizing the issue, which is bound to increase or create conflict within the community. Moreover, such decisions also encourage other actors in the body politic to take positions either in support or opposed to the Court's ruling.

The Public and the Court

In 1837, Alexis de Tocqueville noted:

> The power of the Supreme Court is immense, but it is power springing from opinion. They [the decisions] are all-powerful so long as the people consent to obey the law; they can do nothing when they [the public] scorn it. Now, of all powers, that of opinion is the hardest to use, for it is impossible to say exactly where its limits come. Often it is as dangerous to lag behind as it is to outstrip it.[68]

If the people are not supportive of the Court's actions, the Court will become ineffective and its decisions will be of little consequence. Thomas R. Marshall pointed out in the 1980s that

> [m]ost research on public opinion and judicial policy-making . . . suggests that judges' decisions tend to reflect public opinion—especially when public opinion itself is clearly expressed, one-sided, and intense.[69]

Marshall does acknowledge that there have been important exceptions to this—as, for example, surrounding the issue of school prayer[70]—but in his research, he found that of the 146 case decisions he looked at, decided between 1935 and 1986, "some 62 or 63 percent of the Court's decisions were consistent with the polls when a clear poll majority (or plurality) existed."[71]

Marshall looked at a variety of issues in his study, but one thing was clear, namely, when the environment was one of the issues considered in opinion polls, it rarely fared very well. In a July 2012 *Gallup Poll*, for example, the public was asked to rate different issues by level of importance for the next president to address. In this poll, "dealing with environmental concerns" was ranked by Americans as their lowest priority, while "creating new jobs" received the majority vote of the 1,030 polled.[72]

This has not always been the case, however. The public reflected a more intense interest in the environment during the 1980s and 1990s. This change, Riley Dunlap theorized, could be explained by the negative responses given to these issues by the Reagan and George H. W. Bush administrations, since neither administration inspired the public's confidence that the federal government would do much to protect the environment. It thus was left to the public, and not the elected policymakers, to do what it could to support the environment.[73] As it has been suggested by Stephen Wasby:

> People react to the Court not only in terms of its procedures but also in terms of its results. . . . As it makes policy, the Court interacts with these other branches of government, and public reaction to the Court's decisions (its policy statements) *forces the Court to become an actor in the political system*: if the Court's actions are to have an effect and if, in the long run, the Court is to survive, the justices must take the Court's environment into account [emphasis added].[74]

Will the Future Look Like the Past as Far as the Court and Environment Are Concerned?

One concern about the Court's future effectiveness must be blamed on the Court itself.

Despite all we have said about the Supreme Court's involvement with environmental issues, it seems to have made less of a contribution to the environmental movement than it has to other social issues such as affirmative action. It has given over leadership in many instances to lower courts, Congress, and the president. Richard E. Levy and Robert L. Glicksman supported this idea suggesting that "the Supreme Court appears to have retreated from this activism by emphasizing judicial restraint in its environmental decisions."[75]

However, the Court has made a difference in the environmental area when it has had to decide an issue conflict. An example of this was the case of *TVA v. Hill* (1978),[76] where the Court's reading of the Endangered Species Act left it no choice but to block the Tennessee Valley Authority's

(TVA) building of the Tellico dam project to save the snail darter fish. Here, the Court was supportive of the administration and the Congress and opposed to the wishes of the TVA.

Case Study

The Court and "Those Dam Fish": *TVA v. Hill* (1978)

What do you do with three-inch, bony, perch-like freshwater fish called "snail darters," whose numbers have been estimated between five and twenty thousand, swimming in the Little Tennessee River—the very river where the multimillion-dollar Tellico Dam is being built? Do we have a problem? Yes, I think we do! The difficulty came about when the Supreme Court in 1978 declared that the Little Tennessee River was the snail darter's "critical habitat"—the loss of which would likely reduce their chance of survival. Yes, but what about the dam—which was 80 percent completed at the time the Court made its declaration—that was to provide electricity for some twenty thousand homes, help promote shoreline development, and provide flood control, recreation, and economic development for an area losing its population. And aren't there forty-five other species of darters in the Tennessee River system? Yes there are—and wouldn't you say that was enough darters for any one river system to have to sustain? Ah, but that ignores the fact that there are no other Percina (Imostoma) Tanasi perch-like freshwater fish! Is the Little Tennessee River the only place these Percina (Imostoma) Tanasi darters live? Apparently so, although the Tennessee Valley Authority (TVA) tried unsuccessfully to relocate the population to the Hiwassee River area to allow for the completion of the dam. And the problem became more intense as a six-member majority of the Supreme Court (that is, Chief Justice Burger, along with Associate Justices Brennan, Stewart, White, Marshall, and Stevens) decided in 1978 that the snail darter was "endangered" and indicated that the Endangered Species Act of 1973 was written precisely to protect any species considered endangered. Indeed, survival of a species was to be preferred over every other value—even multimillion dollar projects that were 80 percent completed! The majority of the Supreme Court accepted the District Court's word on this when it stated: "Whether a dam is 50 percent or 90 percent completed is irrelevant in calculating the social and scientific costs attributable to the disappearance of a unique form of life. Courts are ill-equipped to calculate how many dollars must be invested before the value of a dam exceeds that of the endangered species."[1]

Yet two of the three dissenters (Associate Justices Powell and Blackmun), in defending the dam, felt that it was absurd to think that the Endangered Species Act would bring a halt to a nearly completed project. This decision, they stated, would have damaging results over any project, regardless of how important it was, if it was determined that the project threatened the extinction of an endangered species or its habitat. Associate Justice Rehnquist filed his own dissent.

So, should halting this dam be cause for an environmental victory celebration? You must admit that it was an immediate triumph for the bony perch-like creatures in toppling a dam! And this case was rated in a 2001 survey of environmental law professors as one of the five most important judicial decisions shaping environmental law.[2] However, the Court proved it was no real champion of environmental preservation, since the majority was really supporting separation of powers, as they commended the Congress for writing such a precise act as the Endangered Species Act. Their decision said nothing about the environment being preferred over development. The dissenters did reveal themselves as defenders of development over the environment, which should make environmentalists feel even more ill at ease—putting projects over fish—if it helps the economy. Neither perspective bodes well, in the long run, for the environment in the future! And the bony fish? In 1984, those same snail darters were taken from the "endangered" listing and reduced to a "threatened species" category, giving them less protection than they had, while the Little Tennessee River was removed from the "critical habitat" category, further reducing the darters' protection.[3] You could say, I think, that there is absolutely no respect for bony perch-like freshwater fish any more.

Notes

1. *TVA v. Hill*, 437 U.S. 169 (1978).

2. Kenneth M. Murchison, *The Snail Darter Case: TVA versus the Endangered Species Act* (Lawrence: University Press of Kansas, 2007), 194.

3. "Species Accounts," Endangered and Threatened Species of the Southeastern United States (The Red Book), FWS, Region 4 as of November 1992; http://www.fws.gov/ecos/ajax/docs/life_histories/E010.html. Accessed January 30, 2014.

Globalization is also critical in determining the Court's long-term effectiveness. It affects some issues more than others. For example, such issues as global warming, depletion of the ozone layer, overpopulation, trade, concern for chlorofluorocarbons, as well as today's greatest environmental challenge in global climate change, which affects every nation and person in the world. These issue areas recognize no borders, and require international, not regional or even national solutions.[77] Ironically, because these issues do reach global proportions, they extend beyond a national court's power, and search for an international tribunal or forum for decision. This only exacerbates the difficulties for resolution, given the numbers of persons that must agree to a solution.

Case Study

The Climate Crisis is Real, Mr. Limbaugh, REAL!

Who hasn't ever wished for a warmer winter? Unfortunately, the bitter cold we experienced in the eastern and southern parts of the United States during the winter of 2013–14 is only one side effect of climate change; the others are much less appealing, such as the threat that ocean levels will rise, burying a number of coastal cities in the United States, as well as devastating our crop supply by arbitrarily changing seasons. We have also seen water levels changing both in U.S. rivers and in the rain and snowfall levels that sections of the country can expect each year. Many U.S. policymakers and public spokespersons have recognized the potential devastation that can come from doing nothing about climate change, while others have ignored the reality of it, labeling global climate change a "hoax" and "fraud," as did Rush Limbaugh in 2009.[1]

During the Clinton-Gore years in the White House, the administration did in fact recognize the seriousness of climate change, but the Senate refused to support their efforts. The president and vice president did, however, encourage other nations to support the Kyoto Protocol, a global agreement that bound nations to make the effort to reduce emissions. But there have been years when the governing president has done little positive to respond to climate change. During the two terms George W. Bush was president, for example, rather than support the protocol, Bush renounced it, refusing to join the 191 other nations that were beginning to control emissions as prescribed by the protocol.[2]

This raises an important question concerning who should get primary credit for any success the United States has had regarding climate change policy. Should we give that credit to presidents, to the Congress, to states and local jurisdictions, or to the Supreme Court and federal courts? Since our primary interest in this chapter is in the Supreme Court, we need to ask what the Court has done to respond to the threat of climate change.

The Supreme Court's major contribution in responding to climate change undoubtedly came in its 2007 ruling in *Massachusetts et al. v. Environmental Protection Agency.*[3] The Environmental Protection Agency (EPA) had denied having authority to regulate greenhouse gas emissions during the George W. Bush years, but the Supreme Court ruled that the EPA had such authority and was required to regulate emissions of greenhouse gases, including carbon dioxide emissions, which for the first time was labeled by the Court to be a pollutant.

This Massachusetts decision was reinforced in 2011 when the Court held in *American Electric Power v. Connecticut*[4] that any federal regulation of emissions of carbon dioxide would need to come under the Clean Air Act and be negotiated by the EPA, and not through use of the judiciary's "public nuisance" standard. This further established the EPA as the primary actor in all matters related to climate change regulation. Yet two cases in 2014—*EPA v. EME Homer City Generation, L.P., et al.*[5] and *Utility Air Regulatory Group v. EPA, et al.*[6]—while recognizing the important role played by the Environmental Protection Agency, did make clear that it was an agency limited to the authority given it by Congress through the Clean Air Act. Justice Scalia was clear in the *Utility Air Regulatory Group* case that no agency is free to "adopt . . . unreasonable interpretations of statutory provisions and then edit other statutory provisions to mitigate the unreasonableness."[7]

So whom do we thank for standing up against the climate change crisis that has the potential of affecting every nation and individual? Do we thank environmental groups like Friends of the Earth? Do we thank the Congress? Do we thank such presidents as Jimmy Carter, Bill Clinton, and Barack Obama? Or do we thank that Supreme Court and the lower federal courts? Given that the court has never been an institution to lead out on environmental concerns compared to the

Congress and selected presidents, it has played a legitimizing function in its decisions responding to climate change. As a result, it can have a long-lasting effect in guiding future Court decisions, as well as the decisions of the other policymaking bodies.

Notes

1. Rush Limbaugh, "From Climat4e Hoax to Health Care to 'Hope,' Liberalism is Lies," *Rush Limbaugh Show*, November 23, 2009; http://www.rush-limbaugh.com/daily/2009/11/23/from_the_climate_hoax_to_health_care_to_hope_liberalism_is_lies. Accessed January 28, 2014.

2. "Undermining Environmental Law" (editorial), *New York Times*, September 30, 2002; http://www.nytimes.com/2002/09/30/opinion/30MON1.html?pagewan. Accessed September 30, 2002.

3. 549 U.S. 497 (2007).

4. 564 U.S. _____ (2011).

5. 572 U.S ___(2014).

6. 573 U.S. ____(2014).

7. 573 U.S. _____ (2014).

Conclusion

Although the Supreme Court has resisted assuming a leadership role regarding the environment, leaving that up to the president and Congress, the Court has been relatively active in all aspects of environmental decision making. Clearly, the Court remains one of the most important political policymaking bodies at the federal level, rendering its primary contribution to policymaking in a series of top-tier case decisions rendered over the years. These decisions have seen the Court settling environmental disputes that have developed between the president and Congress, as well as settling multistate disagreements concerning environmental issues. The Court has also empowered the Environmental Protection Agency as a primary decision-making body regarding the regulation of climate and pollution. Regarding air and climate control, the Supreme Court has reserved to itself primary authority to determine what "pollution" means and what an "emission" consists of. Moreover, Court decisions on the environment, as in all areas, become difficult to overturn and change. Its decisions thus have a legitimizing and stabilizing effect on other actors in the political system. In addition, by virtue of the fact that the Supreme Court is a *court* within the political system, it legitimates decision making in a way no other agency or institution can equal. Whether the environment is a good fit for the Court or not, only time and its membership will determine.

Websites

U.S. Supreme Court

www.supremecourt.gov

Supreme Court, Court Cases, and Justices of the Supreme Court over time

www.oyez.org

U.S. Supreme Court opinions

www.findlaw.com/casecode/supreme.html

U.S. Supreme Court, recent decisions and legal information

www.law.cornell.edu/supremecourt/text/home

U.S. Supreme Court blog: Symposium Announcements and Calendar Events

www.scotusblog.com

8

The Global Environment

The preceding chapters in this book have addressed environmentalism within the framework of American domestic politics. For example, in the United States there are numerous federal institutions and agencies with jurisdiction over the issue of climate change. While the president has the power to work with the leaders of other countries in the treaty-making process, the United States Senate has the power to ratify the treaty or oppose it. Moreover, in both the House and the Senate legislators can introduce legislation concerning climate change that might reflect the interests of environmental groups or business and industry. The president can then sign the bill or veto it. Among federal agencies, the Council on Environmental Quality is responsible for assessing the impact on the environment resulting from actions of the federal government; the Environmental Protection Agency (EPA) has been involved in numerous studies on the effects of climate change on the planet, especially in coastal areas; the Department of the Interior assesses the impact of climate change on public lands and waters under its jurisdiction; and the Federal Emergency Management Agency (FEMA) is responsible for establishing a national emergency management system to limit personal and physical losses caused by environmental events such as hurricanes, although it lacks the capabilities and resources to implement effective programs in collaboration with subnational governments.[1]

In this chapter, we examine global environmental policy that shows the linkage between the United States and other nations as well as the impact of transboundary environmental issues. Although our primary focus in this book is on the political aspects of environmental affairs, we are also concerned about the science that plays an integral role in bolstering our understanding of the biosphere. At this point, we are confronted with the "science and politics problem." As Sheila Jasanoff frames the issue, the

171

purpose of policymakers is to "harness the collective expertise of the scientific community so as to advance the public interest."[2] However, as Lynton Caldwell explains:

> Science alone cannot save the environment. Political choice is required to translate the findings of the environmental sciences into viable policies. Scientific information, even in its limited present state, is far from being fully utilized in contemporary society.[3]

Or, as Andrew Dessler and Edward Parson put it:

> In addition to the challenges that policy debates pose to science, science also poses hard challenges to policy debates because citizens and politicians are not generally able to make independent judgments of the merits of scientific claims.[4]

Moreover, as James Rosenau has argued, "Rooted in the processes of nature and the responses of nature to human intervention, environmental issues are inescapably embedded in a scientific context."[5] The interface of nature, science, and politics is an important dynamic in the study of global environmental policy as each plays an essential role. Amid the polemics and debates that occur within and between nations about the proper approach to global environmental protection we are reminded that

> [p]oliticians cannot exercise control over environmental outcomes without recourse to scientific findings. They may claims that the findings are not clear cut or remain subject to contradictory interpretations, but they are nonetheless dependent on what the practices of science uncover about the laws of nature.[6]

The global political environment is comprised of some two hundred nation-states, each with its own interests and priorities. Yet the actions of one or a collection of these countries can affect their neighbors as well as the global natural environment. In an effort to address environmental problems on a regional and global scale, nations have been drawn together in regional and international organizations, conferences, and treaties. Moreover, the environment has increasingly become an integral element of the national security debate and is now an essential part of the foreign policy making and diplomatic process. It is to these concerns that we now turn our attention, but we first assess citizen opinion on the global environment.

International Public Opinion about The Global Environment

In chapter 3, we discussed public opinion about the environment among American citizens. Similar to their counterparts in the United States, citizens in other countries are confronted with numerous public policy problems. To what extent is the environment a salient issue for them? In this section, we provide a portrait of the political orientations of citizens in countries reflecting a different level of development about the importance of environmental affairs.

In a 1993 global survey, citizens were asked to identify their country's "most important problem." Citizens in Mexico (29 percent) and Chile (20 percent) were most inclined to indicate that the environment was the most important problem facing their country. In a second tier of countries— Denmark (13 percent), Japan (12 percent), United States (11 percent), Canada (10 percent), Germany (9 percent), Russia (9 percent), South Korea (9 percent)—citizens were less likely to view the environment as the most important problem. The United Kingdom (3 percent) and the Philippines (2 percent) lagged behind, comprising a third tier of countries. As we can see, two decades ago citizen concerns about the importance of the environment ranged from a high of 29 percent to a low of 2 percent. In short, at the most, no more than three out of ten citizens viewed the environment as the most important problem facing their country. Moreover, we find countries across the range of development rating the importance of the environment favorably or not favorably.

Almost two decades later, a new and interesting profile of countries emerged regarding the degree to which citizens in different countries and different levels of development focused on the environment as their country's most important problem. Unlike the 1993 study, citizens in this 2010 study were more tightly bound ranging from a high of 13 percent to a low of 2 percent. Canada (13 percent) and Denmark (10 percent) were most inclined to view the environment as the most important problem facing their country. The remaining nine countries fell into a mixed bag of second-tier countries.

Based on the comparison of the 1993 and 2010 studies in Table 8.1, two important findings resulting from this observation of citizens' attitudes

Table 8.1. Citizens' Views of the Importance of the Environment (Selected Countries)

The Environment as Most Important Problem

Country	1993	Country	2010
Mexico	29%	Canada	13%
Chile	20	Denmark	10
Denmark	13	South Korea	8
Japan	12	Germany	6
United States	11	Mexico	5
Canada	10	Russia	5
Germany	9	United States	4
Russia	9	Japan	4
South Korea	9	United Kingdom	3
United Kingdom	3	Philippines	3
Philippines	2	Chile	2

Source: Adapted from Riley E. Dunlap, George H. Gallup Jr., Alex M. Gallu, Dennis L. Soden, and Brent S. Steel, eds. *Handbook of Global Environmental Policy and Administration* (New York: Marcel-Dekker, 1999), 11; and Tom W. Smith, "Public Attitudes towards Climate Change & Other Environmental Issues Across Time and Countries, 1993–2010," National Opinion Research Center, the University of Chicago, 2010, at www.norc.org/pdfs/public_attitudes_climate_change.pdf. Accessed December 2, 2014.

about the importance of the environment are (1) the overall importance of the environment as the most important problem dropped from 1993 to 2010; and (2) the decline included both advanced and less developed countries. In short, it is important to note that the global community continues to view environmental affairs as less important than other issues.

Global Environmental Issues

Controversies over environmental issues involving wetlands protection, global warming, biodiversity, and endangered species are important examples of the problems associated with the divergent political and economic interpretations of nature and scientific findings. Moreover, environmental threats are not limited to one geographic locale but instead have a cross-national impact on either a regional or international level. As rivers travel through one country to another they carry pollutants from their origin point. Coal-fired utility plants in one country impact neighboring countries in the form of acid rain. Wildlife does not recognize human-made political and legal borders. Ocean dumping and climate change have both a regional and global impact.

Almost two decades ago, in his study of international environmental policy, Lynton Caldwell categorized environmental issues in terms of their "criticality."[7] Issues considered "critical" included endangered species, loss of habitat, expanding human population, loss of forests, and overgrazing pollution of and decrease in fresh water supplies. Issues that were considered "becoming critical" included loss of topsoil, atmospheric pollution and climate change, energy sources and their alternatives, threats to the biogeochemical processes underlying the biosphere, and the impact of large public works and their maintenance on the availability of resources for the future. For the purpose of illustration, we will focus briefly on an issue that we suggest has evolved from a "becoming critical" issue to one that has become a "critical" issue—namely, sea level rise.

Sea Level Rise

The scientific community has informed us through the Intergovernmental Panel on Climate Change that as a result of global warming and climate change, one major consequence is rising seas, which will pose a variety of threats to coastal communities around the globe. In the United States, more than half (52 percent) of Americans inhabit coastal areas that comprise less than 20 percent of the land in the continental United States.[8] These threats will affect residential and commercial properties, not to mention U.S. national security concerns, as rising waters threaten the largest Navy base in the world, located in southeastern Virginia. Human activities (e.g., burning fossil fuels) are major contributors to the warming of the planet. As Orrin Pilkey and Rob Young put it, "[C]limate is getting warmer," and "the [mountain] glaciers will . . . melt and return water to the sea. . . . Currently, nearly all of the world's glacial ice is retreating and thinning, adding water to the oceans."[9]

In 2010, the U.S. National Oceanic and Atmospheric Administration (NOAA) reported in its Arctic Report Card the "continuing decline of ice melt and sea ice and a decrease in glaciers."[10] Of course, not all coastal areas will be affected similarly. The United States will be challenged in terms of the impact on three different coastal areas—the East Coast, the Gulf Coast, and the West Coast. Recent research conducted by scientists at the University of Arizona suggests that by the year 2100, 9 percent of the land area of 180

American cities along the East and Gulf coasts will be inundated, and that major cities from New Orleans to Tampa and from Miami to Virginia Beach may lose approximately 10 percent of their land area[11] (see also chapter 2).

Globally, rising seas will result in environmental, geographical, social, economic, emotional, and political consequences for the populations of coastal countries especially low-lying countries such as Bangladesh. Moreover, island states such as the Maldives face the challenge of watching their country slowly fall victim to the encroaching ocean. As the warming of the planet continues, potential conflicts are on the horizon, including territorial access to and jurisdiction over newly exposed land with profitably energy reserves, competition over fishing rights, and newly formed shipping lanes.

In 2010, the U.S. National Research Council and U.S. National Oceanic and Atmospheric Administration framed the problem of sea level rise for U.S. policymakers in the following way:

> Coastal counties are among the most densely populated areas in the United States—more than one-third of all Americans live near the coast, and activities along or on the ocean contribute more than one trillion dollars to the nation's economy. This intense development of coastal area has increased their vulnerability to sea level rise and storm surges by decreasing the extent of natural buffers and causing accelerating rates of subsidence.[12]

For Americans and citizens living in coastal settings around the globe, the warning signs are clear—a future characterized by rising waters, recurrent flooding, and policymakers engaged in obfuscation and inaction rather than substantive problem solving.

International and Regional Organizations, Conferences, and Treaties

How have state, regional, and international communities responded to the need for protection of the global environment? In discussing global governance and the formation of international regimes, Oran Young found it useful to organize the notion of international environmental regimes into three categories—namely, the international commons, shared natural resources, and transboundary negative externalities.[13] The international commons is comprised of elements in the biosphere in which members of the global community have shared interests, including global warming, biodiversity, stratospheric ozone depletion, and inner as well as outer space, among others. Flora and fauna and underground fossil and mineral resources constitute shared natural resources where these elements cross over national borders. When one country engages in behavior that has an adverse impact on its neighbor(s), transboundary negative externalities occur. For example, when an industrial site in one nation is polluting a river that travels through other countries, it has committed a cross-national water pollution violation.

United Nations Environment Program

One significant attempt to engage the United Nations in playing a critical role in global environmental affairs was the establishment of the United Nations Environment Program (UNEP) in 1972, subsequent to the United Nations Conference on the Human Environment. Its focus is planetary in providing leadership in its responsibility for addressing the United Nations' focus on the global environment.

For more than forty years, despite a lack of sufficient resources, UNEP has been fairly effective in the formation and coordination of international environmental conventions, negotiations, and research.[14] For example, in 1977 an international scientific forum in Washington, D.C., was organized by UNEP to study threats to the ozone layer. It produced a "World Action Plan" and future research agenda; in 1988, it gave the Intergovernmental Panel on Climate Change the task of evaluating global climate change in order to provide data about human impact on the climate; in 1995, it promoted the Global Biodiversity Assessment, a scientific analysis of the relationship between human beings and biodiversity; and more recently in 2010, it engaged in the Millenium Ecosystem Assessment that provided scientists and world leaders new and important information about biodiversity and contemporary threats to flora and fauna.[15] Looking back, we can see the importance of this one of the many institutions that make up the United Nations system. However, as Speth and Haas lamented, "Despite its small size . . . UNEP . . . has been the spark plug that fired the development of modern global environmental governance. But it has lacked the mandate, size, authority, and resources to do the job expected of the world's environmental leader at the international level."[16]

The European Union

In 1992, in order to expand its membership beyond that of the European Community (EC), which had been established in 1957, the Maastricht

Treaty created the European Union (EU). While the EC had been most concerned with economic growth and development and national security, by the early 1970s it had begun to give attention to environmental affairs, beginning with a series of Environmental Action Programs (EAPs) created to harmonize environmental policy on a cross-national basis. As one among the many policy areas under the jurisdiction of the newer EU, the environment gradually became an integral aspect of decision making in the EU. After all, the densely populated European continent was threatened by pollution, a high rate of natural resource consumption, and increasing waste production.[17]

Over the last two decades, the EU has evolved into an important regional and international institution that has maintained a strong commitment to engage in important efforts in support of protecting the regional and global environment. This has been an interesting task since the EU is comprised of twenty-eight member states each with its own history, culture, political system, and stages of development. Environmental policymaking in the EU has been characterized by legislative directives that are binding on member states. At the same time, the member states have varying political motivation, financial resources, and administrative capacity to respond to EU mandates.

However, despite these potential constraints, for instance, the European Union has been a leader in pushing for climate change regulations to curb greenhouse gas emissions significantly by the end of the first half of the twenty-first century. The EU has invested significant funds in support of this goal. In contrast to the United States, which is divided by partisanship and ideology on this issue, the EU has been characterized by an absence of politicization among member states. One could argue that members of the EU view the climate change issue as an "all in this together" regional problem.

The EU has made progress in its overall effort to improve environmental conditions in Europe. At the same time, as the United States and Canada have demonstrated regarding climate change, the EU's attempts to harmonize relations with other states remains a critical factor as it looks to the future and ongoing and new environmental issues.

International Conferences and Agreements

During the post–World War II period, delegates from numerous countries have met to discuss and then find solutions to environmental problems. The character of ecological issues has changed over the years as new threats have emerged while old issues remain a source of contention. Early regional and international agreements tended to focus on wildlife and marine conservation. By the early 1960s, national security and the environment became an issue as the United States, Soviet Union, and the United Kingdom signed the 1963 Limited Nuclear Test Ban Treaty. From the 1970s to the present biodiversity and endangered species have emerged as important global issues as well as stratospheric ozone depletion and global climate change. Between 1972 and 1997, three major international conferences brought together world leaders to address environmental issues.

In 1972, the United Nations sponsored the Conference on the Human Environment, in Stockholm. The conference brought together delegates from both rich and poor nations, in accord with a proposal presented by Sweden in the early 1960s. The Stockholm conference resulted in the adoption of a set of "common principles" that established a framework within which international cooperation could be fostered, with the goal of improving the health of the global environment. A Declaration on the Human Environment was issued that contained the common principles and more than one hundred action plans.[18]

Two decades later, what became known as the Earth Summit in Rio brought together thousands of delegates to discuss environmental conditions near the close of the twentieth century. Expectations about this United Nations Conference on the Environment and Development were perhaps too high. One observer characterized the conference in this way:

> Faced with an agenda of more than one hundred environmental
> policy issues, more than a thousand pages of negotiating texts,
> and the unprecedented security requirements of the assembled
> [participants], the organizers of the summit understandably
> wanted to be remembered for what they overcame politically
> than for what they achieved in policy terms.[19]

Nonetheless, the conference provided a forum in which a common theme—sustainable development—was accepted and several agreements concluded, namely, the Climate Change Convention, the Biodiversity Convention, and Agenda 21, which focused on a "global partnership for sustainable development" and stressed the need for financial assistance from the rich to the poor countries.[20]

Five years later, delegates from industrialized countries met in Kyoto, Japan, to discuss climate-changing emissions of greenhouse gases. The major problem that challenged the participants concerned the level of reduction in the production of greenhouse gases. While the goal was to reduce emissions to 1990 levels, the participants were given different targets. Due to variations in previous emission outputs, modified target levels for the amounts of emission reductions were established as an incentive to encourage governments to sign the agreement. However, in early 2001 newly elected U.S. president George W. Bush withdrew the U.S. commitment to the Kyoto Protocol. For the next dozen years, the United States, along with Canada, has played the role of laggard at international conferences that have convened to address greenhouse gas emissions. Although Barack Obama has used the power of the American presidency to push for national and international initiatives to address global climate change, congressional Republicans and Democrats representing fossil fuel states have challenged his efforts.

The Environment as a Twenty-First Century National Security Policy Issue

The environment can be characterized as the national security issue of the twenty-first century. This position has been adopted by policymakers both domestically in the United States and internationally.[21] During the Clinton presidency, Secretary of State Madeleine Albright stated unequivocally that the environment was a central element of U.S. foreign policy. Regarding the threat posed by climate change, both Ban Ki-moon, Secretary-General of the United Nations, and Thomas Fingar, U.S. Deputy Director of National Intelligence, have argued that global climate change will have national security impacts for the foreseeable future.

The United States remains a most important political actor in regional and international environmental affairs. International environmental policymaking requires multilateral cooperation, but this is influenced by self-interest as different states seek to limit impositions on their actions. For instance, President George H. W. Bush and President George W. Bush both used potential threats to the U.S. economy and jobs as their excuse to limit the U.S. commitment to international initiatives addressing climate change. George H. W. Bush refused to sign the global warming document at the Earth Summit until "mandatory" guidelines were changed to "volunteer" actions by U.S. business and industry, while George W. Bush rejected the Kyoto Protocol altogether. As Paul Harrison puts it:

> U.S. leadership in the international environmental issue area has not been consistent. Sometimes it leads—as in the case of ocean

dumping and stratosphere ozone depletion; at others, it resists action, despite possibly severe consequences—as in the case of climate change. Nevertheless, in looking broadly at international environmental diplomacy in recent decades, one can see a gradual U.S. engagement with the world in an increasingly multilateral approach to environmental protection.[22]

The linkage between environmental protection and national security comprises areas such as war and environmental degradation, access to vital natural resources, air pollution, deforestation, rising seas, and toxic pollution, among others, many of which combine and overlap. For instance, although the 1963 Limited Nuclear Test Ban Treaty was viewed principally as an arms control measure, Theodore Sorenson argued that President John F. Kennedy was genuinely concerned as well about the environmental and public health threat resulting from nuclear explosions.[23] Further, it has been estimated that during the years of U.S. military involvement in Vietnam, 80 percent of forest land and 50 percent of coastal habitat were destroyed due to the use of Agent Orange; and during the Persian Gulf War in 1990 the coastline of the gulf was polluted by oil spills, marine life suffered, and oil wells burned uncontrollably, emitting toxins into the atmosphere.[24]

As global climate change has gained the attention of a variety of players involved in international environmental affairs, its effect on homeland security has become an increasingly important issue in the United States. For instance, in 2010, the U.S. Navy produced a document that indicated that climate change will be a "contributing factor" to future conflicts, and four years later the U.S. Department of Defense published a Climate Change Adaptation Roadmap for the military to employ in dealing with the challenges posed by climate change. As former secretary of defense Chuck Hagel warned:

> Climate change is a global problem. . . . Its impacts do not respect national borders. No nation can deal with it alone. We must work together, building joint capabilities to deal with these emerging threats.[25]

In the foreword to the 2014 Department of Defense Climate Change Adaptation Roadmap, Chuck Hagel, Secretary of Defense under President Obama and former Republican senator from Nebraska, set forth the preeminent purpose of the defense department: "[T]he responsibility of the Department of Defense is the security of our country. That requires thinking ahead and planning for a wide range of contingencies." He continued:

Among the future trends that will impact our national security is climate change. Rising global temperatures, changing agricultural patterns, climbing sea levels and more extreme weather will intensify the challenges of global instability, hunger, and destruction by natural disasters in regions around the globe. . . . While scientists are converging toward consensus on future climate projections, uncertainty remains. But this cannot be an excuse for delaying action.[26]

In short, global climate change is playing a profound role in the decision-making process of the U.S. Department of Defense and the U.S. Navy as they deal with protecting installations, equipment, coastal infrastructure, and supply chains, access to natural resources, and military and/or humanitarian efforts. Having said this, vocal critics in and out of the U.S. government remain political opponents of efforts to address the impact of climate change on U.S. national security.

In this chapter, we have discussed several important issues that have implications for the global environment. In the following guest essay, Christina Slentz discusses solidarity norms and multilateral efforts to address global climate change.

Guest Essay

Solidarity Norms and International Climate Change Cooperation

Christina Slentz
Graduate Programs in International Studies
Old Dominion University

The problem of climate change is undeniably a global issue requiring significant multilateral cooperation if a solution is to be found. Thus, normative features such as common values, trust, and willingness to collaborate will be important elements in solving this dilemma. What explains the contrast among nations that generally find agreement, exhibit comparable levels of economic development, and, for the most part, share similar values, norms, and expectations? Is it possible that isolating such factors might better direct efforts to foster "greener" behavior worldwide?

Based upon the fact that the most capable and pivotal countries to the successful reduction of climate change effects are those least imminently and intensively threatened, this research sets out to examine national tendencies toward altruism and social cohesion, explores how such attitudes correlate with willingness to mitigate harmful behaviors, and observes readiness toward cooperation in the hopes of shedding light upon what compels international climate change behavior modification. In contrast to a priori expectations that measures of solidarity would correlate with those most willing to cooperate as well as the highest levels of greenhouse gas (GHG) mitigation, data from 2000–2010 suggest that greener behavior and willingness to collaborate on a global scale do not necessarily go hand in hand. Rather, although those nations who are characterized as most socially cohesive are likewise most internationally cooperative, they are not ultimately the most successful in terms of mitigation, calling into question broad, "one-size-fits-all" international efforts that fail to hit the normative mark beyond Europe and demanding consideration of regional approaches more in line with local stakeholder expectations.

The Science and Economics Problem: The Politics of Environmentalism among Nation-States

There is much discussed in the literature describing the limitations of green politics. In particular, Kathryn Harrison and McIntosh Sundstrom point out the heavy economic weight carried by international mitigation treaty compliance costs when balanced against "a normative commitment" in favor of green issues over immediate self-interest.[1] Yale's Center for Climate Change communication director Anthony Leiserowitz agrees, observing that simple climate awareness is by and large "a necessary, but insufficient condition to motivate an individual or collective response," conversely arguing public perceptions of risks and dangers "fundamentally compel or constrain political, economic and social action."[2] Complicating the development of risk awareness is the perception of global warming's impacts as temporally and geographically distant from those in developed countries, and more specifically, American perceptions, where 68 percent of American citizens recognize a threat to "people all over the world" and/or "nonhuman nature" while only 13 percent exhibit concern regarding "impacts on themselves, their family or their local community."[3]

On the question of mitigating fossil fuel usage, local perceptions of climate change damage estimates, energy reform costs, and oil consumption amount to strong regional heterogeneity.[4] While China and the United States willingly subsidize oil production, for example, European and African states seek reductions in use at 46 percent and 60 percent, respectively.[5] This regional heterogeneity and lack of perceived risk largely explains U.S. preferences for a "wait and see" approach toward climate change, departing from the "strong multinational preference for a precautionary approach" held by many in developed countries, who similarly perceive the immediate threat of global warming as geographically removed from them but prefer "major action now," thereby demonstrating a normative, beyond-self global commitment.[6] Coming from a previous leader in international environmental collective action—having led the way to the eradication of ozone-depleting chlorofluorocarbons (CFCs)—American hesitancy serves as a heavy blow to hopes of establishing necessary levels of international trust and motivation to overcome the significant counterbalances of individualistic tendencies and economic self-interest.

As a result, we come to a serious divide among the great powers, which typically steer international consensus and frame global expectations for nation-state behavior. Although scientific consensus offered by the International Panel on Climate Change provides sufficient evidence to support serious perceptions of risk and to motivate universal and immediate response, discord and denial persist. As Sussman and Daynes observe, "While the science informs us that global warming and climate change demand increasingly urgent attention, an additional factor that plays a role in climate change policymaking is the 'science and politics' problem," which introduces significant interest-based politicization into the issue.[7] Dessler and Parson describe the challenge of this debate, pointing to the tricky combination of both positive claims (empirical data) and normative claims (values and principles) that have been interwoven into the discourse and have complicated and confounded cooperative approaches to address this existential dilemma.[8] Consequently, "science" has been unable to resolve this problem, complicated by interests and fuzzy perceptions of the global impact, leaving the international community with a normative dispute.

Noting "deep changes required not just to our energy consumption but to the underlying logic of our economic system" and the heavily politicized sentiment of the anti–climate change contingent, Naomi Klein uncovers the fundamental, strongly held economic principle of free-market neoliberalism—heavily based upon consumption and industrialization and often criticized for its negative impact on the environment—as a key issue responsible for triggering the intensity of this normative debate.[9] Fear of losing or undermining this economic ideological preference challenges not only "big business" but also the relatively conservative individualistic identity that has historically accompanied fiscal policies behind highly unregulated privatized economies and reduced governmental market controls. More accurately than framing this dynamic as the "science and politics" problem is perhaps consequently the moniker, the "science and economics" problem.

Analyzing Solidarity Norms in a Comparative Perspective

This analysis is a comparative exploration of the role of solidarity norms in influencing national proclivity toward collective action that focuses on forty-seven countries chosen on the basis of similar levels of development indicated by membership in the OECD and/or Group of Twenty (G-20). These nations were divided into four analytical groupings determined by their demonstrated commitment to cooperative international climate change policy in the form of the Kyoto Protocol. This agreement, linked to the United Nations Framework Convention on Climate Change (UNFCCC), sets a future path for cooperative efforts to reduce global emissions of harmful greenhouse gases (GHGs) with staggered requirements among nations of varying degrees of development, establishing monitoring and reporting mechanisms as well as adaptation funds for those identified as most likely to suffer the harmful effects of climate change and unable to produce their own responses.[10] Graduated future increases in GHG reductions are outlined for each commitment period to soften industry/economic adjustment, and only the most developed countries, the "Annex I" nations, face specific emissions mitigation requirements at this point. Thus, in accordance with the December 2012 Doha Amendment agreeing upon the latest commitments to the Protocol,[11] the four groupings identified in this study are characterized as follows:

- Group ZERO—failure to ratify the treaty

- Group ONE—ratification without specific GHG reduction requirements

- Group TWO—ratification and agreement to participation in Kyoto Protocol Phase One (2008–12) with either emissions cap or reductions

- Group THREE—ratification and agreement to continued participation in Kyoto Protocol Phase Two (2013–20) with emissions reductions

Within each group, comparisons are made between those displaying the strongest measures of "solidarity" and are further compared to World Bank data revealing actual national reduction of carbon emissions per capita from 2000–2010.[12] As the Kyoto Protocol was adopted in 1997 and the "detailed rules for implementation"—the "Marrakesh Accords"—were adopted in 2001 with the aim of an immediate effort to orient mitigation policies upon emission levels of the baseline year 1990, this time frame gauges initial cooperative behavior and serves as a good starting point for evaluation.[13]

Solidarity can be defined as "unity or agreement of feeling or action, especially among individuals with a common interest, mutual support within a group."[14] Applying the idea of solidarity as a normative expectation characterizing a national population, attitudes of "mutual support within a group" and "social cohesion" are well captured by the measure of willingness to pool resources based upon perceived "unity" and "common interest." Because most countries in the sample are democratic societies, with Saudi Arabia being the notable exception, government revenue mechanisms designed to fund and provide common services for the good of the whole via national income tax are generally implemented with the consent of the people. Therefore, average national income tax rates provide an excellent way to quantify notions of solidarity within a population, illuminating a spectrum of tendency ranging from strong feelings of individualism to strong feelings of common welfare or "solidarity." The year chosen for these data is 2012, the year in which the Doha Amendment was instituted.[15] To gauge how this social preference extends to environmentalism, percentages of Green Party representation among most recently voted-in legislative members of national government are used for further illumination.[16] Understanding the environment to be a generally global and therefore transnational issue, attitudes of green solidarity thus express feelings of unity and social cohesion that transcend borders.

With these two measures of solidarity and real carbon emission mitigation per capita, the sample is divided into the four groupings of international cooperative commitment for analysis. Above-average values for solidarity and mitigation are shaded to represent greater tendency. The strength of solidarity norms can be observed as relating to both international environmental cooperation and real carbon emissions mitigation.

Solidarity and Climate Change Response

Results of this analysis offer a mixed bag on the role of solidarity in relation to climate change response and actual per capita carbon reduction. Overall, "solidarity" correlates strongly with commitment to international cooperation, as fully two-thirds of all Group THREE countries, having accepted the most extensive reduction requirements and the longest-term commitment, exhibit above average national income tax rates averaging 39.1 percent, compared to the sample's 36.8 percent. Average Group THREE Green Party representation is also above average at 4.7 percent—well above the total mean score of 3.5 percent. However, while Group THREE produced an above average reduction in emissions from 2000–2010, with a net decrease of 0.1 metric tons per capita, the United States, the sole Group ZERO country based upon its abstention from the Kyoto Protocol, produced a much heftier reduction of 2.6 metric tons per capita, and Group TWO countries, having departed from Kyoto, returned greater average net carbon reductions at an average of 0.7 metric tons (MT) per capita. (See Table 8.2.)

Table 8.2. Summary Chart (Gray shading indicates above sample average levels)

Country	Average Tax Rate 2012	Per Capita Carbon Reduction (MT) 2000–2010	Green Party Data (voted-in seats only)
GROUP ZERO	35.0	2.6	0.0%
GROUP ONE	33.2	–0.9	0.8%
GROUP TWO	31.3	0.7	3.6%
GROUP THREE	39.1	0.1	4.7%
OVERALL AVERAGE	36.8	–0.1	3.5%

A closer examination of each country grouping reveals interesting results. For example, the U.S. average tax rate at 35 percent is not terribly below the overall sample average of 36.8 percent, proving some normative expectation for pooling resources in solidarity within American society. Yet transnational environmental solidarity does not appear to be significant, as very little Green Party presence characterizes the political landscape with absolutely no representation in the legislature. Strikingly however, the United States boasts the second-highest per capita carbon reduction in the study, penultimate to Canada, an original Kyoto participant that withdrew in the midst of political and economic pressure in December 2011.[17]

As Group ONE countries are by treaty parameters lesser developed, their real carbon per capita levels are, unsurprisingly, increasing over this time period, averaging 0.9 MT and correlating with below average income tax rates (33.2 percent) and below average Green Party representation. The two strongest Group ONE states are Israel, the relatively most developed and sole emissions reducer (0.7 MT), and Mexico, exhibiting the smallest increase at 0.1 MT and boasting a significantly above average Green Party representation of 6.7 percent. Of the remaining states, Turkey presents an interesting case as the only treaty-defined "Annex I" state included in this more poorly performing grouping as it is excused from reduction requirements due to its straddling developmentally defined Protocol categories. Although the "inequity" of this exemption undergirds arguments of treaty hesitators and abstainers, the Turkish Ministry of Foreign Affairs points to the disadvantage felt by those outside of internal EU cooperation strategies. The ministry identifies a key missed opportunity for mitigation and adaptation burden sharing: "EC [European Community] member states have rearranged their GHG emission reduction commitments to reach the 8% reduction commitment of the EC with an agreement among themselves. With this rearrangement, [the] UK for example, has taken a commitment to reduce its GHG emission level 12.5% in comparison to 1990 emission levels; Greece on the other hand agreed to increase its emissions up to 25%. At the end the total commitment of countries, who are both EU members and listed in Annex I, remains unchanged."[18] Thus, meaningful participation can be encouraged through regional cooperation although such an influence is not so obviously revealed in the numbers.

Turning attention to Group TWO, those having departed from the Protocol, generally positive mitigation and Green Party presence provide an overall "green" picture—although average tax rates are not very high, at 31.3 percent, the lowest collective result. Thus, some fickle nature may be attached to these national views of international climate cooperation despite strong current indicators of public opinion in favor of green behavior, reflecting the difficulty of achieving sustained environmental policy. As a result, a lack of deeper normative preference for solidarity produces vulnerability to the ebbs and flows of public opinion in favor of climate change cooperation as other national interests compete for support. Such self-interested flip-flopping reduces trust in the international community and, as game theory suggests, encourages "cheating."

The results for the fourth and final group, most easily reviewed in Table 8.3 (page 190), reveal fairly green proclivities among those most committed to international climate change policy cooperation. "Solidarity" levels are high both in terms of tax rates and Green Party presence. In addition, via the concept of burden sharing, this generally regional entity meets its mitigation expectations. Interestingly, Australia, Iceland, Norway, and Switzerland are the only Group THREE countries outside the EU, suggesting internal burden sharing plays a major role in fostering trust and cooperative behavior. Furthermore, with the exception of the case of Australia—a historical Kyoto Protocol "flip-flopper"—the remaining non-EU nations are geographically proximate, which also supports trust and cohesion as history, culture, and economic interdependence powerfully link these nations together.

The problem of a less than stellar reduction observed in terms of per capita carbon decrease, however, remains, begging the question, How much is gained through cooperation? Clearly, the EU is concerned for economic development, supporting a cooperation architecture within which it can shield weaker states. Some factors deserve consideration. Firstly, the "per capita" nature of the carbon mitigation statistic calls attention to population change, bringing to the forefront a significant challenge. The EU growth rate, for example, is estimated at only .21 percent, much lower than the U.S. rate of 0.9 percent and the rate of per capita carbon reduction leader Canada at 0.77 percent.[19] For countries experiencing tremendous growth, such as many African states whose rates hover closer to a 3.0 percent population growth rate, this increase represents a major difficulty in emissions reduction in the midst of development.[20] Moreover, these nations argue, their historical emissions and current carbon footprint pale in comparison to the highly developed countries of the world. Secondly, the Kyoto

Table 8.3. Group THREE (Gray shading indicates above sample average levels)

Country	Average Tax Rate 2012 (Doha)	Per Capita Carbon Reduction (MT) 2000–2010	Green Party Data from most recent legislative election (voted-in seats only)
GROUP THREE			
Australia	45.0	0.3	4.9%
Austria	50.0	0	10.4%
Belgium	50.0	1.3	8.4%
Czech Republic	15.0	1.5	0.0%
Denmark	55.4	0.6	6.7%
Estonia	21.0	−2.6	3.8%
Finland	49.0	−1.4	7.3%
France	45.0	0.4	3.1%
Germany	45.0	1	10.7%
Greece	45.0	0.7	0.0%
Hungary	16.0	0.5	0.0%
Iceland	46.2	1.5	17.5%
Ireland	48.0	2	0.0%
Italy	43.0	1.2	4.7%
Luxembourg	41.3	−2.5	11.7%
Netherlands	52.0	−0.6	6.7%
Norway	47.8	−3.1	0.0%
Poland	32.0	−0.5	0.0%
Portugal	46.5	1.3	7.0%
Slovakia	19.0	0.3	0.0%
Slovenia	41.0	−0.3	0.0%
Spain	52.0	1.4	0.0%
Sweden	56.6	0	7.3%
Switzerland	40.0	0.4	13.8%
UK	50.0	1.3	0.0%
Bulgaria	10.0	−0.6	0.0%
Cyprus	35.0	0.3	0.0%
Latvia	25.0	−0.8	12.2%
Lithuania	15.0	−0.6	3.9%
Malta	35.0	−0.8	0.0%
AVERAGE	39.1	0.1	4.7%

Protocol targets multiple greenhouse gases in addition to CO_2, the primary offender and focus of the World Bank per capita emissions reduction data chosen for this research. These gases may comprise an added measure of reduction for which this work has not accounted. Thirdly, the Protocol also allows for land use, joint implementation projects, and "clean development mechanism" credits that abate or reverse harmful climate change–exacerbating factors and allows these credits to function as reduction equivalents.[21]

In sum, this study reveals strong connections between solidarity and international environmental cooperation, but the bottom line effectiveness of such cooperation is somewhat inconclusive. It is clear that cohesive relationships among likeminded states in close proximity have a positive effect on willingness to cooperate due to the advantages of burden sharing and perceived trust among trading partners. Motivation toward collective action response also arises from those demonstrating the greatest tendencies toward solidarity despite perceptions of threat being spatially and temporally distant, indicating globally minded feelings of unity. Additionally, these countries display longer-term commitment to collective action while even fairly green-minded states with above average Green Party representation fail to maintain lengthy obligations. Therefore, normative preferences for solidarity are not only shown to enable connections between states and have potential to expand circles of committed partners, but also appear to offer the key aspect of sustainability to commitment.

The Environment, International Economics, and Trust: The Green Stag Hunt

The lens of game theory offers a strong framework through which the problem of climate change might be viewed. Rousseau's famous stag hunt captures this environmental dilemma, posing the problem faced by several hunters who must cooperate in order to bring down a stag capable of providing sufficient amounts of the tastiest meat for the group. The "game" arises when a single party is faced with the opportunity to break away from the group to catch a rabbit, only large enough to provide sustenance for himself, subsequently disrupting the stag hunt and ruining the group's collective efforts.[22]

In the case of a "green stag hunt," the "stag" is the preservation of our natural environment such that harmful living conditions do not develop or are, at the very least, abated. Because this objective is a long-term goal, temptation to "cheat" is stronger—it is hard to pass up the possibility of acquiring a few rabbits, or economic advantages that come at the cost of environmentally unfriendly behavior, before the big capture. Pinpointing when this critical moment will occur leaves questionable room for debate that increases temptation to address self-interests while one can still do so. Problematic in the case of climate change, the nature of this hunt is further complicated when the hunting party consists of some who have enjoyed long histories of consuming the venison of the commons while others have not. Producing an agreement among such parties to correct this disparity with future unequal distribution favoring the previously deprived stretches feelings of unity significantly—requiring very strong solidarity norms. Finally, the problem of climate change is such that this green stag hunt is set amid increasing resource scarcity; thus, the opportunity to cheat is presented less and less over time. "Rational actors" are often predicted to preemptively cheat under such circumstances.

In this study, two crucial parties are revealed participating in the green stag hunt dilemma posed by climate change—those willing to commit to the hunt and those who choose to remain outside collective action. The former, the Group THREE countries, privilege cooperation above all else, perhaps even over effectiveness—achieving a minimal improvement, but doing so as a cooperative. The latter, Groups ZERO and TWO, prize their independence, which may conflict with a pro-climate policy position or may coincide with green behavior without tying the homeland to international agreement. As previously indicated, in the cases of the United States (Group ZERO) and Canada (Group TWO), this report reveals greater success in quantifiable CO_2 emissions reductions per capita than that offered by the cooperators (Group THREE). Long-term reliability, however, is not ensured, and feelings of trust are therefore undermined, threatening overall collective achievement.

The European nations, constituting the most cooperative "players" in this game theory analysis, have come to learn through history that their individual national preservation is highly dependent upon their regional as well as international cooperation. The success of this convention over the past seventy years has consequently transformed into a European norm reflected by the statistics captured in this project and consequently interwoven into the architecture of the Kyoto Protocol due to the strong leadership of the EU in its creation. Lacking the same intimate shared history of the Europeans, non-EU states such as the United States and Canada continue to struggle with the competing forces of economics and social norms. For example, geographical proximity is a factor that correlates strongly with cooperation. Thus, such reconstruction could involve independent, regionally based cooperatives, allowing for more tailored approaches to finding solutions, tradeoffs, and burden sharing between nations with already established patterns of coexistence and subsequent levels of trust, and policies based more comfortably upon local/regional normative expectations.

Is a "Green NAFTA" possible? Geographical proximity and trust fostering regional cooperation pose strong prospects for North America. If compiled into a new grouping, the profiles of these three established trading partners are quite promising. (See Table 8.4). If successful environmental cooperation conventions develop reinforcing more supportive attitudes regionally, the possibility of eventually merging into a larger, global compact is significantly enhanced. In fact, a "North American Agreement on Environmental Cooperation" already exists between the United States, Mexico and Canada, providing an initial policy framework on which further cooperation can be constructed.[23]

Table 8.4. The North American States (Gray shading indicates above sampling average levels)

Country	Average Tax Rate 2012 (Doha)	Per Capita Carbon Reduction (MT) 2000–2010	Green Party Data from most recent legislative election (voted-in seats only)
U.S.	35.0	2.6	0.0%
Mexico	30.0	−0.1	6.7%
Canada	29.0	2.8	3.9%
AVERAGE	31.3	1.8	3.5%
SAMPLE AVERAGE	36.8	−0.1	3.5%

Final Thoughts on Solidarity and Future Cooperation

The science and economics problem of climate change presents a significant challenge for international cooperative policies to develop. Although strong solidarity norms are revealed as fostering willingness to pool resources and share burdens on behalf of international environmental policy, sufficient strength of such norms is found predominantly in the European countries, where intense social cohesion on the basis of shared history, economic interdependence, conventional preferences for cooperation as a means of preservation, and geographical proximity unite these states. In more distant, non-European states such as the United States, on the other hand, strong normative preferences for independence and individualism are more generally averse to large-scale international cooperative measures. As conservative Heartland president Joseph Bast bluntly states, "When we look at this issue, we say, this is a recipe for massive increase in government. . . . Before we take this step, let's take another look at the science. . . . Let's not simply accept this as an article of faith."[24] The "science and politics" or "science and economics" problem emerges once more, inspiring efforts to confuse and obfuscate empirical scientific realities out of fear for personal interests.

Yet the United States and Canada, non-Kyoto participants, stand out as having made great environmental strides despite lacking political will for international cooperation. Without such commitment, the question of sustainable green behavior, however, remains for these players in the "green stag hunt" game. If one party defects from the game, more will question their own participation, and overall cooperation will be devastatingly undermined. Currently, the Kyoto Protocol fails to address the normative concerns of many countries, confounding domestic efforts to garner support and threatening a wide base of support. Although emissions-trading schemes attempt to use market approaches to reduce emissions in a "capitalist-friendly" manner, nation-states are still challenged to put their economic priorities behind their environmental concerns. As the threat is temporally and spatially removed from the most critical players in this cooperative effort, approaches that alleviate these concerns must be continuously worked toward within the international community if long-term, sustainable commitments are to be won.

Regional concepts in accordance with local conventions and diverse approaches that give greater independence to countries as they strive toward greener behavior must be considered. Stronger structures to ensure adherence must also be developed to meet the requirements of establishing a cooperative norm. Following World War I, the League of Nations was constructed on largely European sentiment seeking international institutional means of enforcing peace. Weak structures and failure to take into consideration the wider normative preferences of the rest of the world led to failure. The United Nations, on the other hand, was constructed with firmer structures in place and in accordance with a world transformed into greater normative alignment as a result of the global destruction of World War II. The Kyoto Protocol, as it currently stands, is structurally weak, greatly reflective of EU normative inclinations, and therefore chances "missing the boat" with many key international players. The criticality of climate change cannot risk large-scale defection from this existential game. If the UN Framework Convention on Climate Change is to maintain its course alongside the success of the United Nations and not go the way of the failed League of Nations, it must continue to evolve beyond the expectations of Europe and engage with the cultures, norms, and principles that steer overall global behavior.

Notes

1. Kathryn Harrison and Lisa McIntosh Sundstron, eds., *Global Commons, Domestic Decisions: The Comparative Politics of Climate Change* (Cambridge: The MIT Press, 2010), 3.

2. Anthony Leiserowitz, "International Public Opinion, Perception, and Understanding of Global Climate Change," UNDP Human Development Report 2007/2008 (Occasional paper), (New York: UNDP, 2007), 3–4.

3. Ibid., 8, 14.

4. John Hassler and Per Krusell, "Economics and Climate Change: Integrated Assessment in a Multi-Region World," *Journal of the European Economic Association* (October 2012): 975.

5. Ibid. 977.

6. Leiserowitz, "International Public Opinion," 19–20.

7. Glen Sussman and Byron W. Daynes, *U.S. Politics and Climate Change: Science Confronts Policy* (Boulder: Lynne Rienner, 2013), 24.

8. Andrew Dessler and Edward Parson, *The Science and Politics of Global Climate Change: A Guide to the Debate, Second Edition* (Cambridge: Cambridge University Press, 2010), 32–33.

9. Naomi Klein, "Capitalism vs. the Climate," *The Nation*, Nov. 28, 2011, 11–14; http://www.thenation.com/article/164497/capitalism-vs-climate.

10. "Kyoto Protocol," United Nations Framework Convention on Climate Change; http://unfccc.int/kyoto_protocol/items/2830.php.

11. Conference of the Parties, Eighth Session, Doha. "Doha Amendment to the Kyoto Protocol," December 8, 2012; http://treaties.un.org/doc/Publication/CN/2012/CN.718.2012-Eng.pdf.

12. The World Bank, "CO2 Emissions (Metric Tons Per Capita)"; http://data.worldbank.org/indicator/EN.ATM.CO2E.PC.

13. "Kyoto Protocol."

14. "Oxford Dictionaries" (online); http://www.oxforddictionaries.com/us/definition/american_english/solidarity.

15. Tax Rate data drawn from "Individual Income Tax Rates Table," KPMG; http://www.kpmg.com/global/en/services/tax/tax-tools-and-resources/pages/individual-income-tax-rates-table.aspx.

16. Green Party data drawn from The Central Intelligence Agency, "World Factbook"; https://www.cia.gov/library/publications/the-world-factbook/index.html.

17. Ian Austen, "Canada Announces Exit from Kyoto Climate Treaty," *The New York Times*, December 12, 2011; http://www.nytimes.com/2011/12/13/science/earth/canada-leaving-kyoto-protocol-on-climate-change.html?_r=0.

18. Turkish Ministry of Foreign Affairs, "United Nations Framework Convention on Climate Change and the Kyoto Protocol: UNFCCC and Turkey's Position"; http://www.mfa.gov.tr/united-nations-framework-convention-on-climate-change-_unfccc_-and-the-kyoto-protocol.en.mfa. Accessed November 20,2013.

19. CIA, "World Factbook."

20. Ibid.

21. "Kyoto Protocol."

22. GameTheory.net, "Stag Hunt"; http://www.gametheory.net/dictionary/games/StagHunt.html.

23. The Government of Canada, the Government of the United Mexican States and the Government of the United States of America, "North American Agreement on Environmental Cooperation," 1993; http://www.ustr.gov/sites/default/files/naaec.pdf.

24. Klein, "Capitalism vs. the Climate," 18.

Conclusion

Almost four decades ago, in his May 1977 Environmental Message to the Congress, President Jimmy Carter directed the Council on Environmental Quality and the Department of State, along with several other federal agencies, to assess the state of the environment. In his message, the president observed that "[e]nvironmental problems do not stop at national boundaries. In the past decade, we and other nations have come to recognize the urgency of international efforts to protect our common environment."[27] The subsequent study evaluated a variety of global environmental problems and potential consequences. The urgency of these problems was emphasized, as was the need to formulate goals and implement strategies to resolve them before they worsened. The major finding of the 1980 report suggested that:

> If present trends continue, the world in 2000 will be more crowded, more polluted, less stable ecologically, and more vulnerable to disruption than the world we live in now. Serious stresses involving populations, resources, and environment are clearly visible ahead. Despite greater material output, the world's people will be poorer in many ways than they are today.[28]

As we have seen, regional and international cooperation has resulted in numerous agreements for the purpose of improving the quality of the environment. At the same time, while some international organizations have been effective in promoting environmentalism (e.g., United Nations Environment Program) others have been at the center of controversy and criticism for putting material interests above environmental concerns.

Whether and to what extent global environmentalism will have the same impact as the "Renaissance, the Reformation, and the Industrial Age," as one observer argues,[29] remains to be seen. Nonetheless, the human impact on the environment has received increasing attention by governments both rich and poor as well as by citizens around the world. For example, in 1987 political leaders were able to come together to forge an agreement to address the problem of stratospheric ozone depletion. In contrast, global warming and the greenhouse effect continue to be a contentious issue. Political leaders and members of the scientific community argue over what type of action to take with regard to greenhouse gas emissions.

In 2000, a group of scientists working with the Intergovernmental Panel on Climate Change confirmed that for the first time in human history, "an ice-free patch of ocean . . . has opened at the top of the world . . . and is more evidence that global warming may be real."[30] Where Antarctica was the center of attention regarding the "hole in the ozone," issue, the Arctic had not previously assumed an important place in the environmental debate over global warming, the greenhouse effect, greenhouse gas emissions, and global climate change.

During the Bush presidency (2001–09) and the Obama presidency (2009-present) we have experienced what can be called a Tale of Two Presidents. From day one, Bush made it clear that he was not a friend of the environment. The most obvious and salient action taken by the president that illustrated his orientation toward the environment was his rejection of the Kyoto Protocol only two months into his presidency. During his presidency, his administration was characterized by misleading and obfuscating Congress and the American public. A dark chapter during the Bush administration involved political appointees revising scientific reports in order to make environmental threats appear less threatening than they actually were. Midway through his presidency, at the same time that some scientists and lawmakers said the White House was selectively using studies to fit its political agenda, news reports were pointing out that "top scientists and environmentalists accused the Bush administration . . . of suppressing and distorting scientific findings that ran counter to its own policies."[31] In contrast, Obama has included the environment in general and climate change in particular near the center of his policy agenda. Through his appointments to various agencies, especially the Environmental Protection Agency, he has provided a clear indication of the importance of the environment as an important public policy issue. Where Bush was compelled to talk about the United States' addiction to oil, Obama, for instance, used the resources of the White House to persuade auto manufacturers to seek greater fuel efficiency.

President Barack Obama has pushed a green agenda regarding climate change, in particular, during his terms in office. However, he has been consistently opposed by congressional Republicans and congressional Democrats representing fossil fuel states. The same can be said about the role of the United States abroad, where he has attempted to work with global partners and especially with the Canadian government in what could become a North American alliance in support of effective global climate change policymaking. However, domestic constraints have limited his ability to secure a leadership position on the global environment.

In his book *Earth Odyssey* (1998), Mark Hertsgaard relates his experiences traveling around the world during the 1990s to almost twenty different countries to learn more about the health of the global environment and the future of the human species. He concludes his journey of several years with the following admonition:

> The outlook is uncertain, the hour is late, the earth a place of both beauty and despair. The fight for what's right is never ending, but the rewards are immense. Humans may or may not still be able to halt the drift toward ecological disaster, but we will find out only if we rouse ourselves and take common and determined action.[32]

Ten years later, *New York Times* journalist Thomas L. Friedman argued in his book *Hot, Flat, and Crowded* (2008) that the world has a problem:

> It is getting *hot, flat, and crowded*. That is, global warming, the stunning rise of middle classes all over the world, and rapid

population growth have converged in a way that could make our planet dangerously unstable. . . . The hour is late, the stakes couldn't be higher, the project couldn't be harder, the payoff couldn't be greater.[33]

As we head toward the middle of the second decade of the twenty-first century, we are confronted by political observers who are divided on the question of whether and to what extent the United States can and will engage in a positive role regarding the global environment. As Sheila Jasonoff has argued, the goal is "how to harness the collective expertise of the scientific community so as to advance the public interest."[34] As we have seen in the preceding chapters, however, in the American political setting, policymakers are challenged by several factors—organized interests, electoral politics, partisanship, and ideological orientation—that will play a crucial role in promoting or obstructing global environmental policy. In short, as Sussman argues, when the "U.S. assumes a leadership role, it bolsters the international effort to promote global environmental protection. When it fails to provide leadership, it weakens that effort."[35]

Case Study

Biodiversity and Endangered Species

Although there are obvious natural reasons for the demise of animal and plant species, human activities play a large role. For instance, a major impact of an ever-increasing human population on the biosphere concerns the degradation of natural habitats. As the number of people increases, there is the likely expansion into surrounding ecosystems. According to a study by Paul Harrison, there is a direct relationship between any increase in human population and the decrease in wildlife habitat.[1]

In his effort to encourage preservation of the "genetic diversity of the biosphere," Lynton Caldwell made reference to the concept "genocide." Caldwell argued that the "term is customarily applied to the elimination of genetic types among humans. But humanity has been guilty of genocide against a vast number of life forms. Since prehistoric times, men have systematically—if also inadvertently—eliminated species of plants, animals, and ecosystems numbering in uncalculated thousands."[2] Efforts have been made in the United States as well as other countries to address the problem of species loss. For example, during the administration of Franklin D. Roosevelt, one of

the early biodiversity efforts was the creation of migratory bird treaties between the United States and Mexico and the United States and
Canada in order to promote wildlife conservation. Four decades later,
in 1973, the United States Congress passed and President Nixon signed
the Endangered Species Act, which promoted an activist approach to
threatened and endangered species in the country. Moreover, that same
year, the Convention on International Trade in Endangered Species of
Wild Fauna and Flora (CITES) was concluded. CITES was one among
several international agreements in the early 1970s established to preserve the natural environment for future generations.[3]

In a 1998 Biodiversity in the Next Millenium survey of biologists,
botanists, and others in related fields, these scientists raised concerns
about a mass extinction explosion that has not been rivaled in any other
time in history.[4] Seven out of ten of them were concerned that by the
third decade of the twenty-first century, 20 percent of contemporary
species will face extinction, while one out of three scientists hypoth-
esize that perhaps 50 percent of all species face this threat. If forests
are the primary habitat for animal and plant species at the same time
that a majority of the world's forest habitat has been lost, international
cooperation is needed to address this urgent threat.[5] Deforestation in
Brazil is a clear example of this dilemma. In describing the problems
resulting from the relationship between economics, politics, and the
environment in Brazil, G. Tyler Miller explains:

> Brazil is divided geographically into a largely impoverished
> tropical north and a temperate south, where most industry
> and wealth are concentrated. The Amazon basin, which
> covers about one-third of the country's territory, remains
> largely unsettled. This is changing as landless poor migrate
> there, hoping to grow enough food to survive, and as its
> tropical forests are cut down for grazing livestock, timber,
> and mining or are flooded to create large reservoirs for
> hydroelectric dams.[6]

Despite national and international efforts to preserve wildlife and
wildlife habitat, the threat to animal and plant biodiversity continues.
During the summer of 2000, the Convention on Biological Diversity
(a product of the Earth Summit in Rio in 1992) convened in Kenya
to address contemporary and future threats resulting from the impact
of human population on biological diversity and the continuing escala-
tion of species loss. In the face of the anticipated impending demise of
countless animal and plant species, it will require extraordinary political

will among government leaders and citizens alike to reverse the trend steering us toward this loss of biological diversity.

Fourteen years later, what is the status of global biodiversity? According to the International Union for the Conservation of Nature, "With the current biodiversity loss, we are witnessing the greatest extinction crisis since dinosaurs disappeared from our planet 65 million years ago. Not only are these extinctions irreversible but they also pose a serious threat to our health and well-being" due to "habitat loss and degradation," "over-exploitation of natural resources," "pollution and diseases," and "human-induced climate change."[7] To put it bluntly, as a result of poaching in Africa, the death of a protected northern white rhino at the San Diego Zoo Safari Park in December 2014 leaves only five rhinos on the planet.

To put this into perspective, the United Nations Environment Program has informed us that more than "60 percent of the world's people depend directly on plants for their medicines."[8] Yet tropical rainforests that have been referred to as "nature's pharmacy" continue to be threatened by ongoing human activities.

Notes

1. See "Facing the Future: People and the Planet: Extinction of Species"; http://www.facingthefuture.org/environment/enviro-4.htm. Retrieved August 8, 2000.

2. Lynton Keith Caldwell, *International Environmental Policy: From the Twentieth to the Twenty-First Century* (Durham: Duke University Press, 1996), 318.

3. See Edith Brown Weiss, "Intergenerational Equity: Toward an International Legal Framework," in *Global Accord: Environmental Challenges and International Responses*, ed. Nazli Choucri (Cambridge: The MIT Press, 1993), 339.

4. See Ed Ayres, "Worldwatch Report: Fastest Mass Extinction in Earth History"; http://www.enn.com/enn-features-archive/1998/09/091698/fea0916.asp. Retrieved August 8, 2000.

5. See "Governments Sound Biodiversity Alarm in Nairobi"; http://www.enn.com/2000/NATU...22/biodiversity.conference.enn/index.html. Retrieved August 8, 2000.

6. G. Tyler Miller Jr., *Living in the Environment*, 7th ed. (Belmont, CA: Wadsworth, 1992), 225.

7. "Why Is Biodiversity in Crisis?" International Union for the Conservation of Nature, October 14, 2010; www.iuch.org/iyb/about/biodiversity_crisis/. Accessed November 27, 2014.

8. The State of the Planet's Biodiversity," United Nations Environment Program May 22, 2014; www.unep.org/wed/2010/english/biodiversity.asp. Accessed November 27, 2014. See also "Biodiversity Survives Extinctions for Now," *Scientific American*, April 20, 2014; www.scientificamerican.com/podcast/episode/biodiversity-survives-extinctions-for-now/. Accessed November 27, 2014.

Websites

Global Environmental Policy Organizations

CONVENTION ON BIOLOGICAL DIVERSITY

www.biodiv.org

CONVENTION ON TRADE IN ENDANGERED SPECIES

www.cites.org

EUROPEAN ENVIRONMENT AGENCY

www.eea.ua.int

GREENPEACE INTERNATIONAL

www.greenpeace.org

INTERGOVERNMENTAL PANEL ON CLIMATE CHANGE

www.ippc.ch

INTERNATIONAL CHAMBER OF COMMERCE

www.iccwbo.org

INTERNATIONAL WHALING COMMISSION

www.iwc.org

ORGANIZATION FOR ECONOMIC COOPERATION AND DEVELOPMENT

www.oecd.org/env

UNITED NATIONS COMMISSION ON SUSTAINABLE DEVELOPMENT

www.un.org/esa/sustdev/csd.htm

UNITED NATIONS ENVIRONMENT PROGRAMME

www.unep.org

UNITED NATIONS POPULATION FUND

www.unfpa.org

U.S. STATE DEPARTMENT

www.state.gov

WORLD METEOROLOGICAL ORGANIZATION

www.wmo.org

WORLD TRADE ORGANIZATION

www.wto.org

WORLD WILDLIFE FUND

www.worldwildlife.org

9

American Politics and the Environment: Conclusion

The Environment: A Social Issue Lacking in Clarity

Environmental policy is one of the oldest social issues. In fact, Professor Eric H. Cline maintains that "climate change has been leading to global conflict—and even the collapse of civilizations—for more than 3,000 years. Drought and famine encouraged internal rebellions in some societies and sacking of others, as people fleeing hardship at home became conquerors abroad."[1] Because the environmental policy debate has been prominent for many years, it is no surprise that many U.S. citizens see themselves as "environmentalists." Jedediah Purdy estimates that "[m]ore than two-thirds of Americans call themselves environmentalists."[2] Other researchers have estimated the percentage to be even higher.[3] Purdy maintains that those considering themselves "environmentalists" come from various segments of society including "a growing list of corporate executives, some of the country's most extreme radicals, and ordinary people from just about every region, class, and ethnic group."[4] Certainly, every modern president who has incorporated environmental issues into his social agenda would call himself an environmentalist. Yet it is often unclear what people mean when they identify themselves as "environmentalists."

Divided and Indistinct

As an issue with many nuances, there is not one definition of environmentalism. Environmentalism can be divided into at least three separate types: (1) "romantic environmentalists," (2) "managerial environmentalists," and

(3) "environmental justice."[5] John Muir and Teddy Roosevelt, respectively, were the most prominent articulators and environmentalists of note of romantic and managerial environmentalism, which represent approaches that have long distinguished the environmental movement. Their supporters are commonly referred to as preservationists and conservationists.

Each of these three approaches to environmental policy has its own distinct objectives, priorities, and methods of achieving its purposes; this often leads to conflict among environmentalist groups. As an example, Purdy notes that preservationists, motivated by the ideas of John Muir and the Sierra Club, are inspired by their "love of beautiful landscapes," but have little concern for economic needs or public access. Conservationists, on the other hand, value the goals and aspirations of Teddy Roosevelt, who attempted to balance ecological maintenance and economic considerations related to natural resource exploitation. Those advocating environmental justice are set apart by their focus on environmental protection as it relates to human rights and equality.[6] The environmental justice movement began to gain salience in the late 1960s, and had emerged as a major movement by the 1980s. Advocates argued that the adverse results of environmental degradation fell most heavily on African Americans and the poor, who were not in a strong political or economic position in society. Advocates of each of these three approaches to environmental policy may well demand that their distinct approach be given priority in the public forum.[7]

Among modern presidents, Franklin Roosevelt turned the nation's attention to conservation in the United States by aggressively supporting such issues as land management and conservation of natural resources. Bill Clinton emphasized environmental justice by arguing that citizens have the right to a clean and safe environment regardless of where they live and spend their time. Barack Obama has closely followed the pattern set by the Clinton administration, but has been more focused and outspoken on global climate change, which is seemingly the most vexing global environmental crisis of this century. Climate change has led to the current extremes in weather and climate around the world. In fact, Obama's Climate Action Plan states that "the 12 hottest years on record have all come in the last 15 years."[8]

Complex and Diverse

In addition to the lack of focus resulting from the diverse environmental movement strategies, environmentalism is also an issue area of great complexity and diversity. It is the most intricate and segmented of the United States' social issues, influencing more sectors of society than any other social issue. Environmental concerns extend to every level of our federal system—local, state, and national governments—and can also influence the international community. Some environmental issues even affect all of these jurisdictions simultaneously. Acid rain, for example, may be a problem for a particular city, a regional problem for those states affected by the rain, and it can become an international problem, as it reaches across national borders.

The environment's complexity, diversity, and global nature can also make it difficult for both the public and policymakers to easily understand its nature. As we made clear previously in the book, environmental concerns in the 1930s were much easier to grasp because they were focused more narrowly on specific conservation issues and the need to protect such issue areas as national forests, public lands, air, and water. But as Al Gore mentioned in his 1992 book *Earth in the Balance,*[9] environmental concerns have become broader and more complex with the rise of issues such as global warming, overpopulation, ozone depletion, and concerns regarding chlorofluorocarbons.

Because they have neither the time nor the incentive to develop an expertise in the subject area, scientific and technological aspects of environmental legislation lead some congresspersons and judges to spurn environmental issues. Yet congresspersons would do well to remember that such expertise probably does exist in executive branch institutions such as the Environmental Protection Agency and other agencies staffed by the president. If members of Congress expect to have a strong influence on environmental policy they need to write legislation precisely so as to reduce the possibility for discretionary interpretation by other decision makers.

Primary Observations Concerning American Politics and the Environment

Throughout the book we have examined how the American political system has facilitated responses to environmental concerns, and which political decision makers have been most important in shaping policy. We will now outline our concluding observations in a series of propositions:

Proposition #1a: Activist presidents have been, and will continue to be, primary actors in shaping environmental policy.

Despite many policymakers being involved in the creation and implementation of environmental policy, presidents have been and will continue to be the most influential of them. If the environment becomes an integral part of a president's social agenda, and if presidents exert aggressive leadership, they can make a major difference in how government responds to environmental concerns. Theodore Roosevelt, as our first active conservationist president, used his presidential powers and authority, in addition to the authority granted him by the Antiquities Act of 1906, to establish national parks and monuments, and to secure forest lands. Through these efforts he was able to protect more land than any other president.[10] He and Gifford Pinchot, who would later head the U.S. Forest Service, set down a conservationist strategy that later presidents, including Franklin Roosevelt, Richard Nixon, Bill Clinton, and Barack Obama, would follow and expand upon.

Franklin Roosevelt also became enamored with forestry and was able to draw on this interest in responding to the Great Depression. Through his creation of the Civilian Conservation Corps (CCC), the government provided millions of unemployed young men with work planting trees, correcting soil erosion, and protecting wildlife refuges in national parks and forests.

Richard Nixon also showed how important presidential leadership can be as he became a leading force for the environmental movement and the only prominent modern Republican environmentalist. Nixon proposed the establishment of the Environmental Protection Agency, and encouraged the passage of major environmental legislation that became key to later environmental protection. His leadership helped to pass legislation that affected clean water and air, protected open spaces, and established new national parks.

George H. W. Bush, early in his presidency, declared that he too wanted to follow conservationist principles in the tradition of Teddy Roosevelt. His primary contribution to the advancement of environmentalism was his support of amendments to the Clean Air Act (PS 101-549). He was unsuccessful in winning the support of environmentalists in the later years of his presidency, because of his association with the anti-environmental policies of the Reagan presidency when he had been Reagan's vice president.

Bill Clinton was initially seen as a major supporter of environmentalism, but because of partisan conflicts with the Congress, Clinton ended up with a mixed record on environmental policy. While in his first term his contributions were limited, he did bring a number of individuals into his administration with proven environmental experience, including Bruce Babbitt, his secretary of the interior, Vice President Al Gore, and others who were supportive of environmental protections.

During his second term, Clinton made his primary environmental contributions by using the Antiquities Act to set aside land as national monuments, or to expand existing monuments in size, as he did in January 2000.[11] As a result, Bill Clinton preserved more land as national monuments in the forty-eight contiguous states than any other president.[12]

Since Republicans gained control of the House of Representatives in 2010, President Obama has been forced to work independently of the Congress in order to accomplish his environmental goals. Using executive orders, Obama has been able to accomplish much of his energy agenda and to establish a National Ocean Council in 2010. This culminated in the 2013 National Ocean Policy, which coordinated executive agencies' efforts at keeping the ocean healthy and reducing conflicts between those with competing interests over ocean use.[13]

During 2013, President Obama created a major Climate Action Plan that he intended to implement with or without congressional support. As he stated in his Second Inaugural Address in January 2013:

> We the people, still believe that our obligations as Americans are not just to ourselves, but to all posterity. We will respond to the threat of climate change, knowing that the failure to do so would betray our children and future generations. Some may still deny the overwhelming judgment of science, but none can avoid the devastating impact of raging fires and crippling drought and more powerful storms. The path towards sustainable energy sources will be long and sometimes difficult. But America cannot resist this transition, we must lead it. We cannot cede to other nations the technology that will power new jobs and new industries, we must claim its promise. That's how we will maintain our economic vitality and our national treasure—our forests and waterways, our croplands and snow-capped peaks. That is how we will preserve our planet, commanded to our care by God. That's what will lend meaning to the creed our fathers once declared.[14]

President Obama's rather complex action plan has three primary objectives. They are: (1) cutting carbon pollution through stringent new rules, similar to the way the administration has reduced mercury and arsenic levels; (2) preparing the country for the impact of climate change by assisting state and local areas to "strengthen our roads, bridges, and shorelines so we can better protect people's homes, businesses and way of life from severe weather"; and (3) taking the lead internationally on means to respond to global climate change and its potential impacts. He hopes to respond to global climate change by "galvanizing international action to significantly reduce emissions . . . and drive progress through the international negotiations."[15]

A major example of President Obama's commitment to taking action on climate change could recently be observed when the president finished negotiating a major agreement on November 12, 2014, with China regarding the threat of climate change. In the agreement, President Obama and President Xi of China agreed that China would stop the growth of its emissions by 2030 while the United States agreed to emit 26 to 28 percent less carbon by 2025 than it emitted in 2005. As Mark Landler suggests, compared with what the United States had previously agreed to, this new agreement will "double the pace of reduction it [the U.S.] targeted for the period from 2005 to 2020."[16] President Obama was very supportive of this negotiation, suggesting to President Xi that "[w]hen the U.S. and China are able to work together effectively . . . the whole world benefits."[17]

Through the actions of these presidents, one can see how important sympathetic presidential leadership is to the advancement of the environmental movement.

Proposition #1b: Activist presidents who oppose environmental protections have been, and will continue to be, detrimental to the advancement of the environmental movement.

Presidents maintain a unique role in the political system, because they are the most visible of all policymakers and are at the very center of the political system. When presidents have actively opposed the environmental movement, environmental advancement has been decidedly slowed on their watch.

Among the limited number of presidents who might be considered anti-environmental, Ronald Reagan's and George W. Bush's attitudes toward the environment, as well as their use of their appointment power to fill key administration positions, proved especially detrimental. Two of Reagan's appointees—Anne Burford (EPA head) and James Watt (interior secretary)—effectively carried out the wishes of the president in undercutting environmental programs. In addition, environmental budgets devoted to water conservation were significantly reduced in 1985. Reagan also did his best to undercut previous bipartisan support for the environmental movement on noncontroversial matters such as clean air, safe drinking water, and the Superfund.

The Reagan administration was unprecedented in its unsympathetic leadership regarding environmental protection. One positive aspect of

Reagan's anti-environmental administration that might be noted was the growth in environmental interest group membership, as environmentalists and others who were disillusioned with the administration, confronted the administration's antipathy toward environmental concerns. Environmentalists hoped these groups would offer some core resistance to the administration's anti-environmental policies.

Reagan's efforts were damaging to the environment, but because his anti-environmental position proved to run counter to public opinion, he failed, beyond his two terms, to facilitate an anti-environmental direction for presidents who would follow.

George W. Bush encouraged a pro-development, antiregulatory, pro-business agenda. He downplayed issues such as air and water purification, protection of wildlife, conservation, and global climate change. Environmentalists were troubled to see how Bush 43 attempted to undermine many of the environmental advances put forward by President Bill Clinton. A specific example is found in the treatment of wetlands. Clinton, in March 2000, had made an effort to protect streambeds from disruption. Bush relaxed restrictions on those streams that did not flow all year, allowing local officials more authority to conduct surface mining; Bush also refused to protect some twenty million acres of wetlands from the dumping of sewage.[18] President Bush preferred environmental concerns to be governed by state and local decision makers. On the Public Broadcasting System broadcast of the *News Hour* on March 29, 2001, George W. Bush's first sixty days were highlighted. The broadcast noted that he had undercut his own promises on clean air, reversed Clinton's initiatives on drinking water, and had begun to withdraw support from the Kyoto Protocol.[19] Katharine Q. Seelye, in the *New York Times* on November 18, 2001, outlined some of the anti-environmental actions taken during Bush's early years in office. These included allowing roads to be built in national parks and letting snowmobiles into national parks. In addition, she indicated that the Bush administration had encouraged mining companies to exploit public lands, and advocated drilling for oil in Alaska's Arctic National Wildlife Refuge.[20]

Another practice that distressed environmentalists and Democrats alike involved Bush's efforts to introduce legislation that appeared to weaken established environmental laws. His "Clear Skies Initiative," for instance, was designed to replace the Clean Air Act that had been passed during Nixon's years in office, and would allow power plants to buy and sell pollution credits, which allowed companies the right to pollute.[21] Another troubling law Bush introduced was his "Healthy Forests Initiative," which, again, reversed Clinton's policies by exempting millions of acres of national forests from environmental protection and allowing the logging and the sale of timber from old growth forests.

By their lack of support for the environmental protections both presidents Ronald Reagan and George W. Bush slowed the advancement of environmental concerns.

Proposition #2: A pro-environmental Congress is pivotal to the success of the environmental movement.

A Congress that opposes the goals of environmentalism or, at the very least, ignores them, regardless of which party is in control, can prove damaging to the environmental movement. Supportive statutory law is essential to the growth of environmentalism. Without such landmark legislation as the National Environmental Policy Act (1969), the Clean Air Act (1963), and its amendments in 1990, the Safe Drinking Water Act, the Surface Mining Act, the Toxic Substances Control Act, the Endangered Species Act (1966), and the Comprehensive Environmental Responses, Compensation and Liability Act ("Superfund Act"), to mention only a few, the environmental movement would have enjoyed little progress and would have lost its direction.

Strong environmental legislation not only becomes the means for Congress to shape environmental policy, but it also serves as a source for other institutions such as the Supreme Court and the presidency to become involved with the environmental movement. In addition, supportive legislation is able, among other things, to counterbalance anti-environmental efforts by other institutions that might represent active supporters of development and business, as occurred during the Reagan years. Congresspersons in favor of the environment saw legislation of consequence pass during these years despite having a president who was by all measures anti-environmental.

Passage of such landmark environmental legislation is difficult for Congress given the nature of its organization. Its dispersed and decentralized power and fragmented authority make focused leadership difficult to bring to bear. In Congress, there are multiple committees and subcommittees with authority over different aspects of the environment. Michael Kraft, in *Environmental Policy and Politics*, notes that some eleven different committees in the House and Senate have authority and jurisdiction over environmental affairs,[22] and estimates that are even higher have been made by other students of politics, with some suggesting that as many as two-thirds of the standing committees in the House and Senate have some influence on environmental legislation. The latter estimate surely takes account of all of the environmental areas covered by committees and subcommittees, including natural resources, forestry and land management, air and water, fisheries and wildlife, energy and public works.

In addition to legislation, a Congress supportive of the environment can also statutorily expand the powers of administrative agencies such as the Environmental Protection Agency (EPA). This extension of power is another tool Congress uses to shape policy during the implementation stage, since it is agencies such as the EPA that will implement the legislation Congress passes. President Obama has not found a sympathetic Congress to support his environmental initiatives, particularly with regard to the Republican-controlled House of Representatives during the 113th Congress. Obama's inability to gain support for environmental policies has made it necessary to use executive agreements to sidestep congressional opposition. He may find even more opposition facing his environmental policies in his last two years in office, as Republicans successful in the midterm elections of 2014 now have significant control of both the House and the Senate. As an example, Senator James M. Inhofe (R-Oklahoma), one of the primary deniers of climate change, who has compared the Environmental Protection Agency to the Gestapo, assumed the chairmanship of the Senate Environment and Public Works Committee in 2015, which promises to pose additional problems for the president.[23]

Proposition #3: The Supreme Court offers important systematic leadership in setting standards for other policymakers and rendering judgments that facilitate the functioning of the political system.

At first, the Court got into the environmental issue arena reluctantly, with justices expressing concerns about their initial involvement. The elaborate nature of the environment, its complexity and diversity, has at times proven a challenge for the justices, but the court has made some very important and lasting decisions affecting environmental matters. The leadership of the Court on environmental matters is distinct from the other institutions at the federal level. It is in a position to set standards for other policymakers and to render opinions regarding system disputes.

There have been a number of decisions along these lines that have been important for the political processing of the environment. For example, the Supreme Court has consistently strengthened the Environmental Protection Agency (EPA), which has allowed the executive branch to issue environmental protections. This has been particularly true for President Obama, who has faced opposition from the Republican-controlled House of Representatives.

The EPA has often had to take responsibility for new regulations. In the 2007 *Massachusetts v. EPA*[24] case, the Court ruled that greenhouse gases were pollutants and could therefore be regulated by the EPA. In 2011 the Supreme Court, in *American Electric Power Co. v. Connecticut*,[25] ruled that corporations could no longer be sued under common law for greenhouse gas emissions since the EPA now controlled standards for emission.

The Supreme Court has also often decided important authority disparities between the president and the Congress regarding the environment. For example, in in the 1975 case of *Train v. City of New York*[26] the Court sided with the Congress, refusing to allow the president to impound environmental funds without authorization from the legislature.

The Supreme Court has also protected federal agencies, executive departments, and bureaus from having to comply with environmental restrictions. In *Winter v. Natural Resources Defense Council (NRDC)* (2008), the U.S. Navy was allowed to continue training sailors in the use of sonar despite its threat of damage to aquatic mammals.[27]

Furthermore, the Supreme Court has made it a practice to settle interstate disputes regarding the environment. In a 1987 ruling, *International Paper v. Ouelette*,[28] the Court ruled that Vermont's environmental protection laws did not apply to International Paper Company's pollution of a lake that bordered Vermont since the source of the pollution was in New York State.

Conflicts of federalism have also been settled by the court, as it did in the 1935 case of *U.S. v. Oregon*[29] where Oregon contested the use of land but the Supreme Court ruled that the federal government prevailed over the state since the land had previously been used as a federal bird sanctuary.

Thus, we see that the Supreme Court renders significant leadership by setting the policy standards that other decision makers must follow, and by resolving questions of authority regarding environmental policy disputes.

Whether the Court has consistently been eco-friendly in its decisions has depended on who has occupied the bench. There have been justices such as William O. Douglas, Thurgood Marshall, Harry Blackmun, William J. Brennan Jr., John Paul Stevens, Stephen G. Breyer, Ruth Bader Ginsburg, and, more recently, associate justices Elena Kagan and Sonia Sotomayor, who have all been quite supportive of the environment. Those justices who have been more opposed to environmental concerns have included Chief Justice John Roberts, Justices Antonin Scalia, Sandra Day O'Connor, Anthony M. Kennedy, and Lewis F. Powell Jr., Clarence Thomas, and Samuel Alito. But the positions of the justices have often depended on the nature of the questions posed in a particular case, and the nature of the issues raised. Anthony Kennedy and John Roberts have, on occasion, surprised students of the Court by joining the liberal block on the Court in support of the environment. The Court in 2014 has included five justices who have been more opposed to the environmental protections than supportive, but, again, this can change depending on the individual circumstances of each case.

The Supreme Court has always maintained an important role in the political system. Thus, we cannot discount it entirely when talking about the environment. The Court's effectiveness with regard to the environment, however, will always depend on who sits on the Court at any one time.

Proposition #4a: Environmental policy is, by and large, originated at the federal level and implemented at the state and local level.

In areas where salient environmental issues have arisen, citizens tend to prefer that state and local governments control environmental policy, and often see federal regulations and actions as intrusive. Nevertheless, it is at the federal level where environmental policy is most often initiated and managed. In some of the more important pieces of legislation, such as the National Environmental Policy Act (NEPA), it is specified that Congress, the president, and the federal courts will have a deciding role on environmental issues, while little mention is given to state and local jurisdictions. Implementation, on the other hand, is largely left to the state and local governments.

It only makes sense that, given the resources that must be involved in any environmental program, the federal level would be the source of policy initiation. Recognizing the need to solicit state and local support, however, federal decision makers use methods of coercive regulation, as well as nonregulatory means such as collaboration and grant money, to persuade the state and local governments to support these programs.

It is always uncertain, of course, whether states will fully support federal policy. States have varied a great deal in their commitment to the environment. Some states such as California have made an effort to support legislation that would improve the quality of the environment and assure its continued health, whereas other states have opposed any federal effort to improve the environment. Those states most supportive of the environment, as measured by the portion of their budget devoted actions on behalf of the environment, have included Wyoming, Alaska, Wisconsin,

Idaho, Montana, Nevada, South Dakota, Maine, California, and Vermont. States that have notably devoted fewer funds to environmental improvement include North Carolina, Massachusetts, New Jersey, Ohio, and Arizona. It would appear that it is the South that has been the most resistant to supporting environmental legislation.

Proposition 4b: When the federal government ceases to be the originator of environmental policy is more likely to be initiated by the states and local areas.

When the federal government has proven reticent to act, states and local areas have tried to be the governmental source for environmental policy. This was particularly evident during the presidency of George W. Bush, who was more supportive of economic development than he was of pro-environmental policy. During his presidency both the federal government and the EPA were inactive and/or opposed to environmental policy. As a result, some states did make the effort to fill the gap. As Nicholas Lutsey and Daniel Sperling wrote in 2008,

> Over the past decade, the federal and state governments have diverged in their awareness and willingness to act on climate change in the U.S. The balance of environmental federalism has shifted decidedly toward lower-level government action on climate change policy.[30]

In particular, California was a leading state in adopting some of the Kyoto standards at the state level and in encouraging other states to follow its lead. California also used renewable fuel initiatives to cut down the greenhouse gas emissions from light-duty vehicles. Moreover, thirty-one other states adopted mandates to blend biofuels with their transportation fuels. Such states as Minnesota and Hawaii followed California's lead. And even Utah, a traditionally red state, exerted some environmental leadership during George W. Bush's presidency, under the guidance of Salt Lake City's two-term Democratic mayor, Rocky Anderson (2000–08), and Jon Huntsman, moderate Republican governor of Utah from 2005–09, wherein the city adopted environmental standards to respond to its air quality problem. For example, it converted the city's fleet to "low- or no-emission vehicles," adopted "anti-idling ordinances," installed "new rail lines," and increased the number of bike lanes. According to Morgan Jacobsen, the city also plans to phase out two-stroke engines in its maintenance equipment, create a program to replace wood-burning stoves, raise the minimum standard for new and renovated municipal buildings to Leadership in Energy and Environmental Design Gold certification, provide assistance to owners of existing buildings for energy efficiency upgrades, and develop tailpipe emissions standards for city departments.

Furthermore, Jacobsen says, plans are being made to make "Salt Lake City International Airport the most energy-efficient airport terminal in the country over the next five to ten years."[31]

Proposition #5: Public support for the environment is crucial to the environment's success in the political system.

Public opinion has been fairly consistent over time, with more than six of every ten Americans favoring stronger standards of protection for the environment even in times of economic instability.[32] Gallup reports that in the thirty years it has asked respondents about stronger standards of protection for the environment, "Americans have almost always chosen the environment over economic growth as a priority."[33] Yet this has not included all environmental concerns. Climate change is not a threat to the environment for which Americans have a strong concern. Only about 24 percent of Americans report, as of 2014, that they are really worried about climate change. It therefore comes in near the bottom of a listing of fifteen issues Americans rated as a concern in a March 6–9, 2014, Gallup survey. According to Gallup, the economy, federal spending, and health care top the list.[34] The public has been generally committed to protecting the environment over the years, but support has varied over time and with the individual issue in question. In 1965, for example, Opinion Research Corporation found only 28 percent of the public saw air pollution as a serious problem, yet five years later, in 1970, some 69 percent of the people indicated that air pollution was a severe problem.[35]

It is also unclear how strong the public's commitment is when the environment is compared with such issues as economic and political concerns. Paired with other issues, the environment has done less well than when the public is asked to consider the importance of the environment by itself. When the public was asked which issue was the most important facing the country in a January 2013 CNN/ORC poll of 814 adults nationwide, only 2 percent of respondents selected the environment, and 46 percentchose the economy, while 23 percent cited the federal budget deficit.[36] In the same year, when 1,772 persons were asked in the Quinnipiac University Poll which issues they were most interested in hearing President Obama address, the environment only attracted 3 percent of those polled, whereas the economy again was selected by 35 percent and the federal budget deficit by 20 percent of those polled.[37] Interestingly enough, however, when people are asked to indicate the importance of the environment as an issue, without considering other issues, then a sizeable percent of the respondents indicated that they believed it to be important.

Survey results indicated the public's support for environmental policy seems to be subject to both the way the question is worded as well as the particular circumstances at hand. Riley Dunlap points out that public eagerness for environmental protections could be explained by the nature and attitude of the administration in power at the time. It was during the Bush administration, for example, that the public strongly felt that the government ought to be doing more to secure the environment.[38]

Public support of the Supreme Court's involvement in environmental policy is a particularly important determinant of whether a decision will be implemented. As Thomas Marshall indicated, most decisions do reflect public sentiment[39]; when the public is uninformed, however, public opinion is unlikely to have a strong effect on policymaker's decisions. Public "education" largely takes place through radio, television, newspapers, and the Internet, all of which are involved in distributing information. Samuel Hays's conclusion is that environmental concerns become mainstream, "only when environmental information becomes more central in the mainstream media."[40] Unfortunately for the environmental movement, the media tend to cover sensational news stories, a category in which environmental issues usually do not fit.

Proposition #6a: Active pro-environmental groups are essential to making environmental policy more visible and understandable to the public and policymakers.

In addition to the media, the public learns about the environment from information distributed by environmental interest groups. While there are many such groups today that support environmental protection, they differ in their tactics, goals, and strategies. On the political left are such groups as the Friends of the Earth and the Sierra Club—the interest group founded by preservationist John Muir. Other important interest groups in the political system that support the environment include Greenpeace, Defenders of Wildlife, and the National Wildlife Federation, and such groups as the National Audubon Society.

These groups are independent of one another, often reflecting the diversity within the environmental movement itself. The earlier groups such as the Sierra Club, the National Audubon Society, the Wilderness Society, and the National Wildlife Federation, established when environmentalism was more focused on public lands and wildlife, share more in common with one another than with the more recent groups such as Environmental Defense Fund, Friends of the Earth, and Natural Resource Defense Council—groups that tend more to mirror the new environmental interests. Some groups have been able to adjust their focus with the years, adapting to new causes. One such group is the National Wildlife Federation, which used to be most concerned with public lands and wildlife, but now addresses air and water pollution, biodiversity, and ozone depletion.

Groups also differ in the narrowness or breadth of their interest, in their resources, and in their political clout. There are the single-issue interest groups such as Audubon and People for the Ethical Treatment of Animals (PETA), and the "Save the _____" (fill in the animal) groups with very narrow bases, that can focus all of their membership and resources on a limited area. The broader-based groups such as Friends of the Earth and the League of Conservation Voters are more diversified in their efforts and, therefore, need more resources to be effective. Some of the larger groups such as the National Wildlife Federation can rally their larger membership in support of a cause, while smaller groups like PETA must rely on unusual, even shocking, tactics to get their message to the people.[41]

If these groups could come to some kind of consensus, they would be more effective contributors to the environmental movement.

Proposition #6b: Active anti-environmental groups make environmental policy less visible and confuse the public and policymakers.

Here we can add such groups as those active in the fossil fuel industry, the coal industry, manufacturing, mining groups, the U.S. Chamber of Commerce, and organizations that sympathize and give support to senators like James Inhofe (R-Oklahoma). During the midterm election of 2014, many of these groups gave funding to climate change deniers and those supportive of the Keystone Pipeline, a pipeline that was intended to transport oil from Canada to the U.S. Gulf Coast. *The Nation* (November 10, 2014) listed the amounts given by many of these organizations, including:

AT&T: $3,270,708
Koch Industries: $3,054,800
American Bankers Association: $2,984,450
National Automobile Dealers Association: $2,808,000
National Beer Wholesalers Association: $2,661,250
ExxonMobil: $2,058,724
PricewaterhouseCoopers: $1,856,877
Goldman Sachs: $1,757,104
General Electric: $1,756,457

The period covered by these donations included the years from the 2007–08 election cycle through July 2014. The money went to some 160 members of Congress who had denied the scientific link between climate change and pollution.[42] These groups included those in the categories suggested as well as some Wall Street firms, banks, and such service companies as AT&T and General Electric.

Proposition #7: The Democratic Party tends to be more supportive of the environment than the Republican Party.

Over time, there have been differences between the parties over environmental policy in the Congress. This became particularly intense during the Republican-dominated 104th Congress wherein such committees as the Senate Environment and Public Works Committee found the environmental conflict particularly rancorous.[43] This became a problem for Democratic President Bill Clinton. Democrats, overall, have been more supportive of the environment at the federal as well as the state and local level than are Republicans as demonstrated by their party platforms over the years. During the decade prior to 1944, the Democrats focused on overcoming the Depression and fighting a war, which momentarily diverted their attention from environment issues. However, since 1944 Democrats have included some statement of support for the environment in every one of their party platforms. As Art Swift noted in 2014, the Gallup website indicated that "two-thirds of Democrats say the environment should be prioritized higher, while about one-third of Republicans say the same thing. This is the largest partisan gulf since 1997, mainly as a result of the sharp rise among Democrats prioritizing the environment higher than economic growth." Swift pointed out, however, that both parties today give "higher priority to the environment than they did prior to the 2008–2009 economic recession."[44]

Republicans, on the other hand, have not been so predictable, but have varied in their support of the environment depending on the nominee for office. When Richard Nixon ran for office in 1968, there was a platform statement on the environment, but four years earlier, in 1964, with Barry Goldwater the nominee, there was no such statement.

When examining the Congress from the 1970s through the 1990s, it would appear that the Democrats have increased their support for the environment at the same time that Republicans have weakened theirs. The gap between the parties appears to be greatest in the West, and less significant in the South.[45] Congressional leaders in each party reflect an even more extreme position on the environment than do the rank-and-file party members, distancing the parties farther and farther apart. Having said this, of course, each party has members on both sides of the issue. During the long years of service by Senator Robert Byrd (D-WV), he was much less supportive of environmental legislation than was his colleague Senator Tom Harkin (D-IA). The reason for this, of course, was the reliance of West Virginia on coal. In the Republican Party there are also very different attitudes in response to the environment. In the House, for example, Sherwood Boehlert (R-NY) has been very supportive of the environmental movement, while his colleague from California, Richard W. Pombo (R-CA), has been critical of environmental protections.[46] The current senators from West Virginia also show a difference in responding to environmental policy. Senator Joe Manchin III (D-WV) was the most supportive given a score in 2013 by the League of Conservation Voters of 38 percent and a lifetime score of 49 percent.[47] The new senator elected from West Virginia in 2014, Shelley M. Capito, has indicated a very moderate support of the environment consistent with her Republican colleagues.

There exist several differences between the two parties. For example, Democrats tend to seek governmental means for reaching their environmental objectives, whereas Republicans frequently look for market-based methods to resolve their environmental concerns. Thus, many Republicans tend to be less supportive of governmentally sponsored environmental programs such as the Endangered Species Act, which invited attempts by conservative Republicans in 1997 to undercut the act where it addressed flood-control projects. One Republican, Congressman Sherwood Boehlert (R-NY), risked irritating his fellow conservatives, by demanding that the Endangered Species Act apply to the flood-control projects as it does to other projects, and he introduced legislation that would do that very thing.[48]

Partisan differences are not as clear on the Supreme Court as in the Congress, but they do exist. Republican appointees tend to be more opposed to the environment than do Democratic appointees. William O. Douglas and Thurgood Marshall, both Democrats, were the most supportive of the environment of any justices in the past or in the present. Those justices least supportive of the environment were all appointed by Republican presidents including associate justices Scalia, Kennedy, O'Connor, Powell, and Rehnquist. Four of these were appointed by President Reagan, while Justice Powell was appointed by President Nixon. Some of the justices, although appointed by Republicans, became more liberal while on the Court and changed their views on the environment becoming much more pro-environmental. These included Justices Stevens, Blackmun, and Brennan.

Republican opposition in the Congress against environmental policy has made it difficult for President Obama to encourage support for environmental policy. Both President Obama and Secretary of State John Kerry have been strong supporters of environmental policy. Both tend to raise the issue with foreign leaders whenever possible. Obama's particular concern is becoming a global leader in responding to global climate change. Most of his response has had to come from his own powers as president, given Republican opposition in the Congress.

Proposition #8: Response to future environmental concerns must take account of the entire ecosystem and not be confined to political and geographic boundaries, or to outdated strategies.

Marian Chertow and Daniel Esty, among other students of the environment, urge us to begin to think of environmentalism in a greater context—an ecosystem context, if you will. They point out that we can no longer focus our attention only on the most obvious environmental problems in a segmented way,[49] in other words, we can no longer focus just on the "belching smokestacks and orange rivers that fouled the landscape," but that it is now necessary to also consider the "fertilizer runoff from thousands of farms and millions of yards; emissions from gas stations, bakeries, and dry cleaners; and smog produced by tens of millions of motor vehicles."[50] Our piecemeal approaches relying on fragmented law, as suggested by Chertow and Esty, are not adequate for our needs today.[51]

While such a comprehensive consideration of the environment has the advantage of forcing policymakers to think of alternate responses to environmental problems in all of their interconnected facets, it does not help simplify our understanding of environmentalism in the twenty-first century.

But we must understand that the relative simplicity of the 1970s is gone. The critical issues for the new century that were identified by Lynton

Caldwell a number of years ago, namely, endangered species, overpopulation, depletion of forests, overgrazing, water pollution, loss of topsoil, climate change, energy reduction, threats to biogeochemical processes,[52] demand broad responses today and will continue to need those responses in the future. Moreover, many other important environmental threats today are not geographically specific and recognize no borders, such as air and water pollution, acid rain, wildlife depletion, ocean dumping, etc. The unanswered environmental challenges of today suggest that environmental concerns during the twenty-first century will exceed our current remedies.

Final Thoughts

Nor do projections of the future offer much relief. According to the 1982 *Global 2000 Report* to the president, the world in the future will be "more overcrowded, more polluted, less stable ecologically, and more vulnerable to disruption than the world we live in now. Serious stresses involving population, resources, and environment are clearly visible ahead."[53] This has become particularly evident as we have become more of an interconnected global society through trade, travel, and political considerations. One of the greatest crises that we face in the twenty-first century is global warming and global climate change, which respect no borders but promise to become devastating to the earth as a whole, by altering seasons, disrupting agriculture, and enhancing stresses between nations, affecting even national security.

Global climate change is not a crisis that is going to be resolved by any one nation; it is a world crisis that will require a world response, or global remedies. In other words, we may need to rely more and more on responses from the international sector. The problem with that comes in the institutions we have set up to handle international concerns. Today, world environmental problems are given to the United Nations, a body lacking the resources to respond to the problem areas. The UN is strapped for funds, and the United States has not been very sympathetic and supportive of the UN's efforts in this arena recently. Other institutions such as the European Union and the World Trade Organization may prove even less effective than the UN in responding to environmental challenges, since the WTO, for one, is allowed to override national environmental laws.

It is our prediction that unless effective global resolutions are found and supported to respond to the basic needs of our environment, the next twenty years will see conditions become even worse than they are today in terms of overpopulation, pollution, and ecological instability.

That is why we need to begin thinking in interconnected terms, in terms of entire ecosystems, and in ways to facilitate ecosystem remedies. With an ecosystem approach we will at least be able to harness new creativity and come up with cooperative remedies supported by more than one nation that might allow for the survival of humankind—at least in the immediate future.

Notes

Chapter 1. The American Political Setting and the Environment

1. Scott Harper, "Being Green Will Be Profitable," *The Virginian Pilot*, January 20, 2000, M 15.

2. Lydia Saad, "More Americans Still Prioritize Economy over Environment," *The Gallup Poll*, April 3, 2013; www.gallup.com/poll/161594/americans-prioritize-economy-environment.aspx. Accessed April 10, 2014.

3. Art Swift, "Americans Again Pick Environment over Economic Growth," *The Gallup Poll*, March 20, 2014; www.gallup.com/168017/americans-again-pick-environ-ment-economic-growth.aspx. Accessed April 10, 2014.

4. Nancy K. Kubakek and Gary S. Siverman, *Environmental Law*, 3rd ed. (Upper Saddle River, NJ: Prentice-Hall, 2000), 335–36.

5. Mark Landy, Marc C. Roberts, and Stephen R. Thomas, *The Environmental Protection Agency*, expanded ed. (New York and Oxford: Oxford University Press, 1994), 251.

6. Robert Durant, *The Administrative Presidency Revisited: Public Lands, the BLM, and the Reagan Revolution* (Albany: State University of New York Press, 1992), 226–27.

7. Kristina Horn Sheeler, "Christine Todd Whitman, Balance, and Her Legacy at the EPA," *White House Studies* 8 (2008): 151.

8. Ibid., 150.

9. Rosemary O'Leary, "Environmental Policy in the Courts," in *Environmental Policy: New Direction for the Twenty-First Century*, 7th ed., ed. Norman J. Vig and Michael E. Kraft (Washington, DC: CQ Press, 2010), 143.

10. Evan Ringquist, *Environmental Protection at the State Level: Politics and Progress in Controlling Pollution* (Armonk, NY: M. E. Sharpe, 1993); James P. Lester, "Federalism and State Environmental Policy," in *Environmental Politics and Policy: Theories and Evidence*, 2nd ed., ed. James P. Lester (Durham: Duke University Press, 1995), 53–56.

11. See, for instance, Walter Lippman, *Public Opinion* (New York: Harcourt, 1922); George Gallup and Saul Rae, *The Pulse of Democracy* (New York: Simon, 1940).

12. Byron W. Daynes and Glen Sussman, *The American Presidency and the Social Agenda* (Upper Saddle River, NJ: Prentice-Hall, 2001).

13. Lamont C. Hempel, *Environmental Governance: The Global Challenge* (Washington, DC: Island Press, 1996), 16.

14. Phyllis Starkey, "Using Science," in *Greening the Millennium: The New Politics of the Environment,* ed. Michael Jacobs (Oxford: Blackwell, 1997), 123–29.

15. Walter Rosenbaum, *Environmental Politics and Policy,* 3rd ed. (Washington, DC: CQ Press, 1995), 164–65.

16. John Muir, "A Voice for Wilderness," in *The American Environment: Readings in the History of Conservation,* ed. Roderick Nash (Reading, MA: Addison-Wesley, 1968), 73.

17. Samuel P. Hays, *Exploration in Environmental History* (Pittsburgh: University of Pittsburgh Press, 1998), 92.

18. Ibid., 95.

19. See, for example, the following works by Inglehart: *The Silent Revolution: Changing Values and Political Styles among Western Publics* (Princeton: Princeton University Press, 1977); and *Culture Shift in Advanced Industrial Society* (Princeton: Princeton University Press, 1990).

20. Swift, "Americans Again Pick Environment over Economic Growth."

21. Glen Sussman and Mark Andrew Kelso, "Environmental Priorities and the President as Legislative Leader," in *The Environmental Presidency,* ed. Dennis L. Soden (Albany: State University of New York Press, 1999), 115–16.

22. Ibid., 116. See also Samuel P. Hays, *Conservation and the Gospel of Efficiency: The Progressive Conservation Movement, 1890–1920* (Cambridge: Harvard University Press, 1959).

23. Gifford Pinchot, "Ends and Means," in *The American Environment: Readings in The History of Conservation,* ed. Roderick Nash (Reading, MA: Addison-Wesley, 1968), 59.

24. Jacqueline Vaughn Switzer with Gary Bryner, *Environmental Politics: Domestic and Global Dimensions,* 2nd ed. (New York: St. Martin's Press, 1998), 6.

25. Andrea Gerlak and Patrick J. McGovern, "The Twentieth Century: Progressivism, Prosperity, and Crisis," in *The Environmental Presidency,* ed. Dennis L. Soden, (Albany: State University of New York Press, 1999), 67.

26. Ibid.

27. A. L. Owen, *Conservation under FDR* (New York: Praeger, 1983), 146.

28. President Franklin D. Roosevelt, quoted in "The Civilian Conservation Corps," in *The American Environment: Readings in the History of Conservation,* ed. Roderick Nash (Reading, MA: Addison-Wesley, 1968), 128.

29. Carolyn Long, Michael Cabral, Brooks Vandivort, "The Chief Environmental Diplomat," in *The Environmental Presidency,* ed. Dennis L. Soden (Albany: State University of New York Press 1999), 195.

30. President Dwight D. Eisenhower, *Public Papers of the Presidents of the United States: Dwight D. Eisenhower* (Washington, DC: U.S. Government Printing Office, 1954), 208–209.

31. James Sundquist, *Politics and Policy: The Eisenhower, Kennedy, and Johnson Years* (Washington, DC: The Brookings Institution, 1968), 323.

32. Rachel Carson, *Silent Spring* (Boston: Houghton Mifflin, 1962), 13.

33. John Whitaker, *Striking a Balance: Environment and Natural Resources Policy in the Nixon-Ford Years* (Washington, DC: American Enterprise Institute for Public Policy Research, 1976), 52–56.

34. Lettie McSpadden, "The Courts and Environmental Policy," in *Environmental Politics and Policy: Theories and Evidence*, 2nd ed., ed. James P. Lester (Durham and London: Duke University Press, 1995), 245–46.

35. Ringquist, *Environmental Protection at the State Level*, 43–45.

36. Glen Sussman and Byron W. Daynes, *White House Politics and the Environment: Franklin D. Roosevelt to George W. Bush* (College Station: Texas A&M University Press, 2010), 197.

37. Kubakek and Silverman, *Environmental Law*, 335–36.

38. Secretary of State John Kerry, "Remarks on Climate Change, Jakarta, Indonesia"; at www.state.gov/secretary/remarks/2014/02/221704.htm. Accessed April 10, 2014.

39. Michael E. Kraft, *Environmental Policy and Politics: Toward the Twenty-First Century* (New York: HarperCollins, 1996), 14.

Chapter 2. American Federalism and Environmental Politics

1. Robert V. Percival, "Environmental Federalism: Historical Roots and Contemporary Models," *Maryland Law Review* 54, no. 4 (1995): 1141–81. This historical summary relies heavily on Percival's framework and analysis supplemented by Jacqueline V. Switzer with Gary Bryner, *Environmental Politics: Domestic and Global Dimensions* (New York: St. Martin's Press, 1998) and Richard N. L. Andrews, *Managing the Environment, Managing Ourselves* (New Haven: Yale University Press, 1999).

2. Switzer with Bryner, *Environmental Politics: Domestic and Global Dimensions*, 5–6.

3. Denise Scheberle, *Federalism and Environmental Policy: Trust and the Politics of Implementation* (Washington, DC: Georgetown University Press, 1997), 13–14; and Denise Scheberle, "Partners in Policymaking: Forging Effective Federal-State Relations," *Environment* 40, no. 10 (December 1998): 14–15.

4. Elaine.C. Kamarck, *The End of Government as We Know It: Making Public Policy Work* (Boulder: Lynne Rienner, 2007); and Cass R. Sunstein, *Simpler: The Future of Government* (New York: Simon and Schuster, 2013).

5. Oliver A. Houck and Michael Roland, "Federalism in Wetlands Regulation: A Consideration of Delegation Under Section 404 and Related Programs to the States," *Maryland Law Review* 54, no. 4 (1995): 1243–1313; David M. Hedge and Michael J. Scicchitano, "Regulating in Space and Time: The Case of Regulatory Federalism," *The Journal of Politics* 56, no. 1 (February 1994): 134–53.

6. Lia Parisien, Layne Piper, and Becca Merrifield, "State Innovations: Agencies Enhance Operations and Outreach," *ECOS green report*: Environmental Council of the States (December 2013): 3, 17.

7. Council of State Governments, *Resource Guide to State Environmental Management*, 5th ed. (Lexington, KY: Council of State Governments, 1999), 13.

8. David R. Hodas, "Enforcement of Environmental Law in a Triangular Federal System: Can Three Not Be a Crowd When Enforcement Authority Is Shared by the United States, the States, and Their Citizens?," *Maryland Law Review* 54, no. 4 (1995): 1580; Barry G. Rabe, "Power to the States: Promise and Pitfalls," in *Environmental*

Policy, ed. Norman J. Vig and Michael E. Kraft (Washington, DC: CQ Press, 2000), 42.

9. "2013 Draft Report to Congress on the Benefits and Costs of Federal Regulations and Agency Compliance with the Unfunded Mandates Reform Act," Federal Reports; http://www.whitehouse.gov/sites/default/files/omb/inforeg/2013_cb/draft_ 2013_cost_benefit_report.pdf. Accessed September 20, 2014.

10. Carole J. Cimitile, Victoria S. Kennedy, Henry W. Lambright, Rosemary O'Leary, and Paul Weiland, "Balancing Rise and Finance: The Challenge of Implementing Unfunded Environmental Mandates," *Public Administration Review* 57, no. 1 (January/February 1997): 65–74.

11. 2013 Draft Report, 7, 17.

12. Ibid., 4, 10.

13. Kenneth A. Zarker and Robert L. Kerr, "Pollution Prevention Through Performance-Based Initiatives and Regulations in the United States," *Journal of Cleaner Production* 16, no. 6 (April 2008): 673–85.

14. Ibid., 682.

15. Sandford F. Borins, *Innovating with Integrity: How Local Heroes Are Transforming American Government* (Washington, DC: Georgetown University Press, 1998), 193. See also Jeremy B. Hockenstein, Robert N. Stavins, and Bradley W. Whitehead, "Crafting the Next Generation of Market-Based Tools," *Environment* 39, no. 4 (May 1997): 12.

16. John C. Morris, "The Distributional Impacts of Privatization in National Water-Quality Policy," *Journal of Politics* 59, no. 1 (February 1997): 56–72.

17. Barry G. Rabe, "Racing to the Top, the Bottom, or the Middle of the Pack," in *Environmental Policy*, ed. Norman J. Vig and Michael E. Kraft (Washington, DC: CQ Press, 2013), 36.

Chapter 3. Public Opinion, Interest Groups, and the Environment

1. Riley E. Dunlap, "Public Opinion and Environmental Policy," in *Environmental Politics and Policy*, ed. James P. Lester, 2nd ed. (Durham: Duke University Press, 1995), 108.

2. Anthony Downs, "Up and Down with Ecology—The 'Issue Attention Cycle,'" *Public Interest* 28 (1972): 38–50.

3. Adapted from Swift, "Americans Again Pick Environment over Economic Growth."

4. Ibid.

5. Ibid.; Saad, "More Americans Still Prioritize Economy Over Environment."

6. Rebecca Riffkin, "Climate Change Not a Top Worry in U.S.," March 12, 2014; www.gallup.com/poll/167843/climate-change-not-top-worry.apsx. Accessed March 18, 2014.

7. Joy Wilke and Frank Newport, "Democrats and Republicans Differ on Top Priorities for Gov't," January 28, 2014; www.gallup.com/167084/democrats-republicans-differ-top-priorities-gov.aspx. Accessed April 20, 2014.

8. Jeffrey M. Jones, "Worry About U.S. Water, Air Pollution at Historical Lows," April 13 2012; www.gallup.com/poll/153875/Worry-Water-Air-Pollution-Historical-Lows.aspx. Accessed April 20, 2014; Lydia Saad, "Water Pollution Americans' Top Concern," March 25, 2009; www.gallup.com/poll/117079/Water-Pollution-Americans-Top-Green-Concern.aspx. Accessed April 20, 2014; Frank Newport, "Americans Show Low Levels of Concern on Global Warming," April 4, 2014; www.gallup.com/poll/168236/americans-show-low-levels-concern-global-warming.aspx. Accessed April 27, 2014.

9. Brendan Moore and Stafford Nichols, "Americans Still Favor Energy Conservation over Production," April 2, 2014; www.gallup.com/poll/168176/americans-favor-energy-conservation-production.aspx. Accessed April 30, 2014.

10. Frank Newport, "Nearly Half in U.S. Say Gov't Environmental Efforts Lacking," April 1, 2003; www.gallup.com/poll/161579/nearly-half-say-gov-environmental-efforts-lacking.aspx. Accessed April 20, 2014.

11. Ibid.

12. Moore and Nichols, "Americans Still Favor Energy Conservation over Production." See also Environmental News Service, "Poll: Support for EPA Clean Air Rules Crosses Party Lines," October 12, 2011; ens-newswire.com/2011/10/13/poll-support-for-epa-clean-air-rules-crosses-party-lines/. Accessed May 1, 2014.

13. Gina-Marie Cheeseman, "Majority of Americans Agree: Protecting the Environment Creates Jobs," May 23, 2012; www.enn.com/top_stories/article/44445/print. Accessed May 1, 2014.

14. Environmental News Service, "Poll: Americans Back Federal Subsidies for Renewables, Not Fossil Fuels," November 14, 2011; ens-newswire.com/2011/11/14/poll-americans-back-federal-subsidies-for-renewables-not-fossil-fuels/. Accessed May 1, 2014.

15. Christopher J. Bosso, "After the Movement: Environmental Activism in the 1990s," in *Environmental Policy in the 1990s: Toward a New Agenda* ed. Norman J. Vig and Michael E. Kraft, 2nd ed. (Washington, DC: CQ Press, 1994), 35.

16. Samuel H. Barnes and Max Kaase, eds., *Political Action: Mass Participation in Five Western Democracies* (Beverly Hills: Sage, 1979).

17. Charles C. Euchner, *Extraordinary Politics: How Protest and Dissent Are Challenging American Democracy* (Boulder: Westview, 1996), 20–21.

18. Thomas C. Beierle, "Public Participation in Environmental Decisions: An Evaluation Framework Using Social Goals," Discussion Paper 99-06 (Washington, DC: Resources for the Future, November 1998), 2; www.rff.org/disc_papers/PDF_files/9906.pdf. Retrieved December 15, 1999.

19. Ibid.

20. Jerrell Richer, "Green Giving: An Analysis of Contributions to Major U.S. Environmental Groups," Discussion Paper 95-39. (Washington, DC: Resources for the Future, September 1995); www.rff.org/disc_papers/PDF_files/9539.pdf. Retrieved December 15, 1999.

21. Robert Cameron Mitchell, Angela G. Mertig, and Riley E. Dunlap, "Twenty Years of Environmental Mobilization," in *American Environmentalism: The U.S. Environmental Movement, 1970–1990* ed. Riley E. Dunlap and Angela G. Mertig (New York: Taylor and Francis, 1992), 15.

22. See, for example, Rochelle L. Stanfield, "Environmental Lobby's Changing of the Guard Is Part of Movement's Evolution," *National Journal* 17, June 8, 1985.

23. U.S. Environmental Protection Agency, "National Top 100," October 27, 2014; www.epa.gov/greenpower/toplists/top100.htm.

24. Friends of the Earth et al., *Ronald Reagan and the American Environment: An Indictment, Alternate Budget Proposal, and Citizen's Guide to Action* (San Francisco: Brick House, 1982).

25. Jay Hair, quoted in "November 94 Had a Green Tint Wildlife Federation Finds," *The Washington Post,* December 24, 1994, A4.

26. See, for example, the Sierra Club's website at: www.sierraclub.org and the Wilderness Society's website at: www.wilderness.org.

27. See the National Wildlife Federation's website at: www.nwf.org.

28. "The Birth of Environmentalism"; www.edfr.org/AboutUS/f_birthof.html. Accessed March 10, 2000.

29. Benjamin Kline, *First Along the River: A Brief History of the U.S. Environment Movement,* 2nd ed. (San Francisco: Acada Books, 2000), 90.

30. Zachary A. Smith, *The Environmental Policy Paradox,* 2nd ed. (Englewood Cliffs, New Jersey: Prentice-Hall, 1995), 17.

31. Craig A. Rimmerman, *The New Citizenship: Unconventional Politics, Activism, and Service* (Boulder: Westview, 1997), 66.

32. "Recognizing an Anti-Environment Ruse," *National Wildlife,* October/November 1992, 30; and Thomas A. Lewis, "You Can't Judge a Group by Its Covered," *National Wildlife* October/November 1992, 9.

33. Jane Hoffman and Michael Hoffman, "What Is Greenwashing?" *Scientific American,* April 1, 2009; www.scientificamerican.com/article/green-washing-green-energy-hoffman/?print=true; Devika Kewalramani and Richard J. Sobelsohn, "Greenwashing: Deceptive Business Claims of 'Eco-Friendliness,' " *Forbes,* March 20, 2012; www.forbes.com/sites/realspin/2012/03/20/greenwashing-deceptive-business-claims-of-eco-friendliness.

34. Reported in Jacqueline Vaughn Switzer, *Green Backlash: The History and Politics of Environmental Opposition in the U.S.* (Boulder: Lynne Rienner, 1997), 105.

Chapter 4. Congress, the Legislative Process, and the Environment

1. Joy A. Clay, "Congressional Government," in *International Encyclopedia of Public Policy and Administration,* Vol. 1, ed. Jay M. Shafritz (Boulder: Westview, 1999), 497–98.

2. James E. Anderson, *Public Policymaking,* 7th ed. (Boston: Cenage, 2010); Thomas R. Dye, *Understanding Public Policy,* 14th ed. (New York: Pearson, 2012); B. Guy Peters, *American Public Policy: Promise and Performance,* 9th ed. (Washington, DC: CQ Press, 2013).

3. David H. Davis, *American Environmental Politics* (Chicago: Nelson-Hall, 1998), 23.

4. Jacqueline Vaughn, *Environmental Politics: Domestic and Global Dimensions*, 6th ed. (Boston: Cenage, 2011).

5. Woodrow Wilson, *Congressional Government* (Boston: Houghton, Mifflin, 1885; reprint John Hopkins University Press, 1981), 82.

6. Kraft, *Environmental Policy and Politics*.

7. Theodore Lowi, *The End of Liberalism*, 40th anniversary ed. (New York: W. W. Norton, 2009); Randall P. Ripley and Grace A. Franklin, *Congress, the Bureaucracy, and Public Policy*, 5th ed. (Pacific Grove, CA: Brooks Cole, 1991).

8. Kraft, *Environmental Policy and Politics*, 14.

9. Philip Shabecoff, *A Fierce Green Fire: The Environmental Movement* (Washington, DC: Island Press, 2003), 246.

10. Mary Cooper, "Environmental Priorities: The Issues," *CQ Outlook*, June 5, 1999, 9.

11. Charles Lindblom, "The Science of 'Muddling Through,'" *Public Administration Review* 19, no. 2 (1959): 79–88; Dye, *Understanding Public Policy*; Peters, *American Public Policy*.

12. Michael Lyons, "Political Self-Interest and U.S. Environmental Policy," *Natural Resources Journal* 39, no. 2 (1999): 271–94; Daniel J. Fiorino, *Making Environmental Policy* (Berkeley: University of California Press, 1995).

13. David L. Feldman, Jean H. Peretz, and Barbara D. Jendrucko, "Policy Gridlock in Waste Management: Balancing Federal and State Concerns," *Policy Studies Journal* 22, no. 4 (1994): 589–605.

14. Michael E. Kraft, "Environmental Policy in Congress: From Consensus to Gridlock," in *Environmental Policy*, ed. Norman J. Vig and Michael E. Kraft (Washington, DC: CQ Press, 2000), 124–25.

15. Sussman and Kelso, "Environmental Priorities and the President as Legislative Leader," 118.

16. Thomas Mann and Norman Ornstein, *It's Even Worse than It Looks* (New York: Basic Books, 2013), 172–76.

17. Henry M. Jackson, "Environmental Policy and the Congress," *Public Administration Review* 28, no. 4 (1968): 305; Vaughn, *Environmental Politics: Domestic and Global Dimensions*; Walter A. Rosenbaum, *Environmental Politics and Policy*, 9th ed. (Washington, DC: CQ Press), 91–91.

18. Kraft, *Environmental Policy and Politics*.

19. Fiorino, *Making Environmental Policy*, 63–64.

20. Ibid., 67–68.

21. Elise S. Jones and Will Callaway, "Neutral Bystander, Intrusive Micromanager, or Useful Catalyst? The Role of Congress in Effecting Change within the Forest Service," *Policy Studies Review* 23, no. 2 (1995): 337–50.

22. R. Shep Melnick, "Pollution Deadlines and Coalition for Failure," in *Environmental Politics: Public Costs, Private Rewards*, ed. Michael S. Greve and Fred L. Smith Jr. (New York: Praeger, 1992).

23. Rosenbaum, *Environmental Politics and Policy*, 95–97.

24. Denis Scheberle, *Federalism and Environmental Policy and the Politics of Implementation*, 2nd ed. (Washington, DC: Georgetown University Press, 2004), 3.

25. Rosenbaum, *Environmental Politics and Policy*, 104; Jones and Callaway, "Neutral Bystander, Intrusive Micromanager," 337–50.

26. Fiorino, *Making Environmental Policy*, 63.

27. http://congress.scorecardlcu.org/leadership.htm and http://www.vote-smart.org/index.phtml.

28. Davis, *American Environmental Politics*, 18.

29. Richard Conniff, "The Political History of Cap and Trade," *Smithsonian Magazine*, August 2009; http://www.smithsonianmag.com/air/the-political-history-of-cap-and-trade. Accessed October 30, 2014; Jim Snyder and Martin Christopher, "Politics May Sour Cap-and-Trade Sweeteners in Obama Plan," *Bloomberg*, June 3, 2014; http://www.bloomberg.com/news/2014-06-03/politics-may-sour-cap-and-trade-sweeteners-in-obama-plan.html. Accessed September 29, 2014.

30. Arthur Bentley, *The Process of Government* (Chicago: University of Chicago Press, 1908); David Truman, *The Governmental Process* (New York: Knopf, 1951); Robert A. Dahl, *Who Governs?* (New Haven: Yale University Press, 1961).

31. Rosenbaum, *Environmental Politics and Policy*, 106–107.

32. Zachary A. Smith, *The Environmental Paradox* (Englewood Cliffs, NJ: Prentice-Hall, 1992), 43.

33. David R. Mayhew, *Congress: The Electoral Connection* (New Haven: Yale University Press, 1974), 32–61; Lyons, "Political Self-Interest," 284.

34. Mayhew, *Congress: The Electoral Connection*, 14–18.

35. Lyons, "Political Self-Interest," 285.

36. Ibid., 275.

37. Sharon Buccino, "Public Demand Will Spur Congress," *The Environmental Forum* 15, no. 3 (1998): 38.

38. Vicky Monks, "Capital Games," *National Wildlife* 34, no. 3 (1996): 25.

39. Jonathan Z. Cannon, "If Not in This Congress, Then a Future One," *The Environmental Forum* 15, no. 3 (1998), 39.

40. Mayhew, *Congress: The Electoral Connection*, 132; Lyons, "Political Self-Interest," 287.

41. Mayhew, *Congress: The Electoral Connection*; Morris P. Fiorina, *Congress: Keystone of the Washington Establishment* (New Haven: Yale University Press, 1977); *Environmental Policy and Politics*, 11.

42. Morris P. Fiorina and Paul E. Peterson, *The New American Democracy* (Needham Heights, MA: Allyn and Bacon, 1998), 604; Murray Edelman, *Symbolic Uses of Politics* (Urbana: University of Illinois Press, 1964).

43. Rosenbaum, *Environmental Politics and Policy*, 94, 260, 180; Kraft, *Environmental Policies and Politics*, 58.

44. Rosenbaum, *Environmental Politics and Policy*, 95.

45. Ibid.

46. Richard A. Merrill, "Congress as Scientist," *The Environmental Forum* 11, no. 1 (1994): 20.

47. Fiorino, *Making Environmental Policy*, 35.

48. James Q. Wilson, *American Government*, 5th ed. (Boston: Houghton Mifflin, 1997), 433–37.

49. Ibid.; R. Douglas Arnold, *The Logic of Congressional Action* (New Haven: Yale University Press, 1990); Kraft, *Environmental Policy and Politics*, 58–59; Vaughn, *Environmental Politics*; Lyons, "Political Self-Interest," 271–94.

50. Richard N. L. Andrews, *Managing the Environment, Managing Ourselves* (New Haven: Yale University Press, 1999), 6–7.

51. Marian R. Chertow and Daniel C. Esty, "Environmental Policy: The Next Generation," *Issues in Science and Technology* 14, no. 1 (1997): 77.

Chapter 5. The Environmental Presidency

1. Richard Lowitt, "Conservation, Policy On," in *Encyclopedia of the American Presidency*, ed. Leonard W. Levy and Louis Fisher, 4 vols. (New York: Simon and Schuster, 1994), 1: 289.

2. Kraft, *Environmental Policy and Politics*, 71.

3. The discussion of the theoretical framework has been taken from Raymond Tatalovich and Byron W. Daynes, *Presidential Power in the United States* (Monterey, CA: Brooks/Cole, 1984).

4. Since, as of this writing, all of our presidents have been men, the masculine pronoun is used in this chapter. In addition, references to the presidency and presidents in general also use the masculine pronoun. Let us be quite clear, however, this usage in no way excludes the possibility or anticipation that in the future women may well occupy this position. In fact, former first lady, and secretary of state and Democratic senator from New York, Hilary Clinton, is currently the leading candidate seeking the Democratic Party nomination to become the first woman president of the United States in 2016.

5. Gallup Poll, "In U.S., Fewer Mention Economic Issues as Top Problem," March 7–10, 2013; http://www.g"llup.com/poll/161342/fewer-mention-economic-issues-top-problem."spx. Accessed February 22, 2014.

6. Congressional Quarterly, *Nixon: The First Year of His Presidency* (Washington, DC: Congressional Quarterly Press, 1970), 2.

7. Stanley I. Kutler, *The Wars of Watergate: The Last Crisis of Richard Nixon* (New York: Alfred Knopf, 1990), 78.

8. Congressional Quarterly, *Nixon*, 3.

9. Mary Etta Cook and Roger H. Davidson, "Deferral Politics: Congressional Decision Making on Environmental Issues in the 1980s," in *Public Policy and the Natural Environment*, ed. Helen M. Ingram and R. Kenneth Godwin (Greenwich, CT: JAI Press, 1985), 48.

10. Bill Clinton, "Remarks on the Observance of Earth Day, April 21, 1994," *Public Papers of the Presidents of the United States: William J. Clinton* (Washington, DC: U.S. Government Printing Office, 1995), 744.

11. "Full Transcript: Obama's 2014 State of the Union address," *The Washington Post,* January 28, 2014; http://www.washingtonpost.com/politics/full-text-of-obamas-2014-state-of-the-union-address/2014/01/28/e0c93358-887f-11e3-a5bd-844629433ba3_story.html. Accessed January 29, 2014.

12. Raymond Tatalovich and Byron W. Daynes, eds., *Social Regulatory Policy: Moral Controversies in American Politics* (Boulder: Westview, 1988), 122.

13. Two early studies that support this idea are Lawrence H. Chamberlain, *The President, Congress, and Legislation* (New York: Columbia University Press, 1946); and Ronald C. Moe and Steven C. Teal," Congress as Policy-Maker: A Necessary Reappraisal," *Political Science Quarterly*, September 1970, 443–70.

14. Roger Biles, *A New Deal for the American People* (DeKalb: Northern Illinois University Press, 1991), 34–45.

15. Ibid.

16. Russell Train, "The Environmental Record of the Nixon Administration," *Presidential Studies Quarterly* 26, no. 1 (Winter 1996): 185–96.

17. Kraft, *Environmental Policy and Politics*, 71–74.

18. Dennis L. Soden and Brent S. Steel, "Evaluating the Environmental Presidency," in *The Environmental Presidency*, ed. Dennis L. Soden (Albany: State University of New York Press, 1999), 315–16.

19. Cecil Andrus and Joel Connelly, "Lessons of the Land: How Carter's Interior Secretary Won a Compromise in the Great Alaska Lands Debate"; http://www.adn.com/weak/wearkive/we981108.htm. Accessed November 8, 1999.

20. The basic Nuclear Waste Policy Act (PL 97-425) of 1983 did pass the Congress, but the amendments did not.

21. Paul Rauber, "Elephant Graveyard," *Sierra* 81, no. 3 (May/June 1996), 24.

22. League of Conservation Voters, "National Environmental Scorecard: Overview of the 2013 Scorecard"; http://scorec"rd.lcv.org/overview-2013-scorec"rd. Accessed September 10, 2014.

23. Ibid.

24. See the following sources: *Congressional Quarterly's Guide to the Presidency* (Washington, DC: Congressional Quarterly, 1989), 451; *Congressional Quarterly Weekly Report*, December 19, 1992, 3925–26; and *Weekly Compilation of Presidential Documents 1995–96* (Washington, DC: U.S. Government Printing Office).

25. Office of the Press Secretary, "Statement by the Press Secretary," October 22, 1999. The White House Publications-Admin"Pub.Pub.WhiteHouse.Gov. Accessed October 22, 1999.

26. H.R. 1495, Water Resources Development Act of 2007; http://georgewbush-whitehouse.archives.gov/news/releases/2007/11/200711023.html. Accessed March 1, 2014.

27. Charles M. Cameron, *The Presidential Veto*, May 2010, 4; http;//www.princeton.edu/~ccameron/The%20Presidential%20Veto%20v3.pdf. Accessed March 11, 2014.

28. "House Sustains Carter Public Works Veto," in *CQ Almanac 1978*, 34th ed. (Washington, DC: Congressional Quarterly, 1979); 154–61; http://libr"ry.cqpress.com/cq"lm"n"c/cq"l78-1237956. Accessed March 20, 2014.

29. While Clinton's percentages were much lower, Clinton's overall budget, of course, was much higher. In 1994 Clinton devoted 1.3 percent to the environment, whereas in 1995, he designated 1.4 percent of the budget for environmental pursuits. For Roosevelt, the first-term budget figures were based on funding for the Departments of Agriculture, Interior and the Federal Power Commission, whereas the Clinton figures

came from environmental funds granted to the Departments of Energy, Interior, and Agriculture.

30. U.S. President, Annual budget message to the Congress, fiscal year 1972; January 29, 1971. *Public Papers of the Presidents of the United States* (Washington, DC: Office of the Federal Register, National Archives and Records Service, 1972), Richard Nixon, 1971, 46–68.

31. U.S. President, Annual budget message to the Congress, fiscal year 1974; January 29, 1993. *Public Papers of the Presidents of the United States* (Washington, DC: Office of the Federal Register, National Archives and Records Service, 1975), Richard Nixon, 1973, 32–48.

32. U.S. President, Letter to the Speaker of the House of Representatives and the President of the Senate on Soil and Water Conservation; March 22, 1985, *Public Papers of the Presidents of the United States* (Washington, DC: Office of the Federal Register, National Archives and Records Service, 1986), Ronald Reagan, 1985, 426.

33. U.S. President, Statement of signing the Los Padres Condor Range and River Protection Act; June 19, 1992, *Public Papers of the Presidents of the United States: William J. Clinton*, 1992 (Washington, DC: Office of the Federal Register, National Archives and Records Service, 1993), 985–86.

34. See a splendid listing of significant legislation in Sussman and Kelso, "Environmental Priorities and the President as Legislative Leader," 134–35.

35. See Congressional Quarterly, senate.gov, house.gov in Francine Kiefer, "Will They Ever Work Together?" *Christian Science Monitor Weekly*, March 3, 2014, 22.

36. See "Breaking News: Climate Pollution Bill Passes House," *Environmental Defense Action Fund,* March 6, 2014, email from Sam Parry of the Environmental Defense Action Fund, takeactionedf.org.

37. U.S. President. "Three Essentials for Unemployment Relief (CCC, FERA, PWA), March 21, 1933." *Public Papers and Addresses of Franklin D. Roosevelt* (New York: Macmillan Company, 1938), 80–84.

38. For budget statistic see "EPA's Budget and Spending," USEPA. United States Environmental Protection Agency. 2014-03-04; http://www2.epa.gov/planandbudget/budget. Accessed April 18, 2014. For information on number of persons working at EPA see the following website: www.EP.gov.

39. U.S. President. "Special Message to the Congress about Reorganization Plans to Establish the Environmental Protection Agency and the National Oceanic and Atmospheric Administration, July 9, 1970." *Public Papers of the Presidents of the United States* (Washington, DC: Office of the Federal Register, National Archives and Records Service, 1971), Richard Nixon, 1970, 578–80.

40. U.S. President. "Remarks on Earth Day, April 21, 1993." *Public papers of the presidents of the United States* (Washington, DC: Office of the Federal Register, National Archives and Records Service, 1994), William J. Clinton, 1993, 468–72.

41. "Call for an Investigation into Activities of the Environmental Protection Agency," *Congressional Record—House*, 104th Congress, 1st Session, May 9, 1995, pp. H4609–H4616.

42. Donald F. Kettl, "Did Gore Reinvent Government? A Progress Report," *New York Times*, September 6, 1994, A19.

43. Richard Nathan, *The Administrative Presidency* (New York: Wiley, 1983).

44. Karen O'Connor and Lee Epstein, "Rebalancing the Scales of Justice," *Harvard Journal of Law and Public Policy* (Fall 1984): 483–506.

45. See Mark Dowie, "Friends of Earth—or Bill? The Selling (Out) of the Greens," *The Nation*, April 18, 1994, 514.

46. U.S. President. "Nomination for Posts at the Department of State, May 6, 1993." *Public Papers of the Presidents of the United States* (Washington, DC: Office of the Federal Register, National Archives and Records Service, 1994), William J. Clinton, 1993, 468–72.

47. On July 20, 1995, for example, President Clinton announced his intention to appoint Eileen B. Claussen as the Assistant Secretary for Oceans, Environment, and International Scientific Affairs for the State Department. "President Names Eileen Claussen to Serve as Assistant Secretary of State for Oceans, Environment, and International Scientific Affairs," Office of the Press Secretary, The White House, July 20, 1995, Almanac Information Server, 19; 2 pages.

48. See Glen Sussman and Byron W. Daynes, "An Early Assessment of President George W. Bush and the Environment" in *George W. Bush: Evaluating the President at Midterm*, ed. Bryan Hilliard, Tom Landsford, and Robert P. Watson (Albany: State University of New York Press, 2004), 59.

49. See Sally Jewell's background at http://online.wsj.com/news/"rticles/SB100014 2412788732459090457887742620198954. Accessed March 7, 2014.

50. U.S. President, "A Typical Executive Order [No. 6910] on Withdrawal of Public Lands to be Used for Conservation and Development of Natural Resources, November 26, 1934," *Public Papers and Addresses of Franklin D. Roosevelt* (New York: Macmillan, 1938), Franklin D. Roosevelt, 1934, 477–79.

51. Federal Register, "Executive Order 12291—Federal regulation," National Archives;   http://www.archives.gov/feder"l-register/codification/executive-order/12291.html. Accessed April 23, 2014.

52. Office of the President, "Memorandum on Clean Water Protection," May 29, 1999, *Public Papers of the Presidents of the United States,* Vol. 1 (Washington, DC: Office of the Federal Register, National Archives and Records Service, 2000), William J. Clinton, 1999, 857.

53. Executive Orders: J. Q. Adams-Obama, "The American Presidency Project"; http://www.presidency.ucsb.edu/executive_orders.php. Accessed April 30, 2014.

54. Executive Order 13406 of June 23, 2006, entitled "Protecting the Property Rights of the American People," *Federal Register*, vol. 71, no. 124, June 28, 2006, 36973–74.

55. Peniel E. Joseph, "Obama Goes Solo in his 2nd Term," *History News Network (HNN)*, March 19, 2014; http://hnn.us/article/155038. Accessed March 19, 2014.

56. Executive Order 13514—Federal Leadership in Environmental, Energy, and Economic Performance; http://www.presidency.ucsb.edu/ws/index.php?pid'86728. Accessed March 19, 2014.

57. See Executive Order 13514—Federal Leadership in Environmental, Energy, and Economic Performance; http://www.presidency.ucsb.edu/ws/index.php?pid'86728. Accessed March 19, 2014.

58. "Statement on Signing the Montreal Protocol on Ozone-Depleting Substances, April 5; 1988," *Public Papers of the Presidents of the United States: Ronald Reagan* (Washington, DC: U.S. Government Printing Office, 1990), 420–21.

59. "Remarks by the President and Prime Minister Brian Mulroney of Canada at the Air Quality Signing Ceremony in Ottawa, March 13, 1991," *Public Papers of the Presidents of the United States: George Bush* (Washington, DC: U.S. Government Printing Office, 1991), 254–57.

60. Ibid., 295.

61. "Obama and Harper Establish Clean Energy Dialogue," 2009: http://www.ens-newswire.com/ens/feb2009/2009-02-19-02.asp. Accessed June 3, 2009.

62. USTR. 2014. "Fact sheet: WTO Environmental Goods Agreement: Promoting Made-In-America Clean Technology Exports, Green Growth and Jobs." Office of the United States Trade Representative, July 8; http://www.ustr.gov/"bove-us/press-office/f"ct-sheets/2014/July/WTO-EG"-Promoting-M"de-in-"meric"-Cle"n-Technology-Esports-Green-Growth-Jobs. Accessed October 16, 2014.

63. EPA. 2014. "2014 North American Amendment Proposal to Address HFCs under the Montreal Protocol." Environmental Protection Agency; http://www.ep".gov/ozone/intpol/mp"greement.html. Accessed October 16, 2014.

64. Norman J. Vig and Michael E. Kraft, eds., *Environmental Policy: New Directions for the Twenty-First Century,* 4th ed. (Washington, DC: Congressional Quarterly Press, 2000), 101.

65. Long, Cabral, and Vandivort, "The Chief Environmental Diplomat," 211.

Chapter 6. Executive Agencies and the Environment

1. "A-Z Index of U.S. Government Departments and Agencies," USA.gov, last modified August 22, 2012; http://www.usa.gov/Agencies/Agency-Index.pdf. Accessed November 18, 2014; Dennis Vilorio, "Career Outlook: Working for the Federal Government: Part 1," *United States Department of Labor: Bureau of Labor Statistics,* September 2014; http://www.bls.gov/careeroutlook/2014/article/federal-work-part-1.htm. Accessed September 25, 2014.

2. Ralph P. Hummel, "Bureaucracy," in *Defining Public Administration: Selections from the International Encyclopedia of Public Policy and Administration,* ed Jay M. Shafritz (Boulder: Westview, 2000), 121–27.

3. Benjamin Ginsberg, Theodore J. Lowi, and Margaret Weir, *We the People: An Introduction to American Politics* (New York: Norton, 2011), 523.

4. Hummel, "Bureaucracy"; Max Weber, *Economy and Society,* Volume 1, ed. Guenther Roth and Claus Wittich, trans. Ephraim Fischoff et al. (Los Angeles: University of California Press, 2013), 956–57.

5. Paul C. Light, *The New Public Service* (Washington, DC: Brookings Institution, 1999).

6. The president lacks the authority to alter the jurisdiction or structure of cabinet-level departments or independent regulatory commission, but he is authorized under the Reorganization Act to submit a reorganization plan to Congress, transferring

functions among agencies. Congress, in turn, must approve the plan within ninety days for it to become effective. Within the Executive Office of the President, the president is authorized to reorganize the staff agencies.

7. Kenneth F. Warren, "Adjudication," in *International Encyclopedia of Public Policy and Administration*, vol. 1, ed. Jay M. Shafritz (Boulder: Westview, 1999), 24–25.

8. Raymond W. Cox III, "Administrative Discretion: The Conundrum Wrapped in an Enigma," in *Public Management and Governance: The Seventh Winelands Conference*, ed. J. van Baalen, E Schwell, and A Burger(Stellenbosch, South Africa: University of Stellenbosch, 2000), 249–63; *Federalist #10*; Weber, *Economy and Society*.

9. Ibid.

10. Ginsberg et al., *We the People*, 526–28.

11. "The Cabinet," The White House; http://www.whitehouse.gov/administration/cabinet. Accessed November 18, 2014.

12. Mark Kelso, "Presidents and Environmental Policy: Republican 'Demons' and Democratic 'Angels'?," unpublished paper delivered at the annual meeting of the Western Political Science Association, San Francisco, March 1996, 15.

13. "Department of the Interior: The Budget for Fiscal Year 2015," Department of the Interior; http://www.whitehouse.gov/sites/default/files/omb/budget/fy2015/assets/interior.pdf. Accessed November 18, 2014; "Employment: June 2014," Office of Personnel Management, last modified September 24, 2014; http://www.fedscope.opm.gov/ibmcognos/cgi-bin/cognosisapi.dll. Accessed November 18, 2014.

14. "Fact Sheet on the BLM's Management of Livestock Grazing," U.S. Department of the Interior: Bureau of Land Management, last modified March 28, 2014; http://www.blm.gov/wo/st/en/prog/grazing.html. Accessed November 18, 2014.

15. "Department of Energy: The Budget for Fiscal Year 2015," Department of the Interior; http://www.whitehouse.gov/sites/default/files/omb/budget/fy2015/assets/energy.pdf. Accessed November 18, 2014; "Employment: June 2014," Office of Personnel Management, last modified September 24, 2014; http://www.fedscope.opm.gov/ibm-cognos/cgi-bin/cognosisapi.dll. Accessed November 18, 2014; "About Us," Department of Energy: The Office of Management; http://www.energy.gov/management/about-us. Accessed November 18, 2014.

16. "Congressional Budget Justification: Fiscal Year 2015," U.S. Nuclear Regulatory Commission, last modified July 9, 2014; http://www.nrc.gov/reading-rm/doc-collections/nuregs/staff/sr1100/v30/. Accessed November 18, 2014.

17. "Congressional Budget Submission: Fiscal Year 2015," Executive Office of the President; http://www.whitehouse.gov//sites/default/files/docs/2015-eop-budget_03132014.pdf. Accessed November 18, 2014.

18. Participation varies depending on whether rule making is formal or informal. See Nancy K. Kubasek, Bartley A. Brennan, and M. Neil Browne, *Legal Environment of Business* (Upper Saddle River, NJ: Prentice-Hall, 2009), 509–11.

19. Cornelius M. Kerwin, "Rulemaking," in *Encyclopedia of Public Administration and Public Policy*, vol. 2, ed. Jack Rabin (New York: Marcel Dekker, 2003), 1071–75.

20. Warren, "Adjudication," 26.

21. Douglas Carter, *Power in Washington* (New York: Random House, 1964), 17.

22. Fiorina and Peterson, *The New American Democracy*, 431–32.; Frank R. Baumgartner and Bryan D. Jones, "Agenda Dynamics and Policy Subsystems," *Journal of*

Politics 53, no. 4 (November 1991): 1044–74.; Seong-Ho Lim, "Changing Jurisdictional Boundaries in Congressional Oversight of Nuclear Regulation: Impact of Public Salience" (Paper presented at the annual meeting of the American Political Science Association, 1992).

23. Stephen J. Wayne, G. Calvin Mackenzie, David M. O'Brien, and Richard L. Cole, *The Politics of American Government* (New York: St. Martin's/Worth, 1999), 636.

24. D. Scott Slocombe, "Implementing Ecosystem-Based Management," *BioScience* 43, no. 9 (October 1993): 612–622.; Melissa A. Schilling and Martin Shultz, "Improving the Organization of the Environmental Management: Ecosystem Management, External Interdependencies, and Agency Structures," *Public Productivity and Management Review* 21, no. 3 (March 1998): 293–308.

25. Quoted in Kline, *First Along the River*, 104.

26. Michael E. Kraft, *Environmental Policy and Politics*, 5th ed. (New York: Longman, 2011), 117; Susan Welch, John Gruhl, Susan M. Rigdon, and Sue Thomas, *Understanding American Government*, 13th ed. (Boston: Cenage Learning, 2012), 569.

27. Kraft, *Environmental Policy and Politics*, 139.

28. "Table 32-1 Federal Budget by Agency and Account," Office of Management and Budget; http://www.whitehouse.gov/sites/default/files/omb/budget/fy2014/assets/32_1.pdf. Accessed November 18, 2014; "Employment: June 2014," Office of Personnel Management, last modified September 24, 2014; http://www.fedscope.opm.gov/ibmcognos/cgi-bin/cognosisapi.dll. Accessed November 18, 2014.

29. Kraft, *Environmental Policy and Politics*, 139–40.

30. "FY 2014–2018 EPA Strategic Plan," Environmental Protection Agency, last modified April 10, 2014; http://www2.epa.gov/sites/production/files/2014-09/documents/epa_strategic_plan_fy14-18.pdf. Accessed November 26, 2014; see also Walter A. Rosenbaum, "Science, Politics, and Policy at the EPA," in Norman J. Vig and Michael E. Kraft, *Environmental Policy: New Directions for the Twenty-First Century*, 8th ed. (Washington, DC: CQ Press, 2012), 163–64,175.; Richard N. L. Andrews, *Managing the Environment, Managing Ourselves* (New Haven: Yale University Press, 2006), 251–52.

31. EPA, "FY 2014–2018 EPA Strategic Plan."

32. Robert Durant, *Why Public Service Matters: Public Managers, Public Policy, and Democracy* (New York: Palgrave Macmillan, 2014), 17.

33. *Performance and Purpose: Constituents Rate Government Agencies* (Washington, DC: Pew Research Center for the People and the Press, April 2000).

34. Ibid., 7, 8.

35. Rosenbaum, *Environmental Politics and Policy*, 91–92.

36. David Vogel, *Fluctuating Fortunes: The Political Power of Business in America* (Washington, DC: Beard Books, 2003).

37. "Beyond Red vs. Blue: The Political Typology," *Pew Research Center*, June 2014: 136; http://www.people-press.org/files/2014/06/6-26-14-Political-Typology-release1.pdf. Accessed December 1, 2014.

38. "Evaluating Government Agencies and American Institutions," *CBS News Poll*, October 2014, 8; http://www.cbsnews.com/news/cbs-news-poll-confidence-cdc-nosedives-since-ebola/. Accessed December 1, 2014.

39. "June 15, 2014 Poll Results," *Wall Street Journal/NBC News Poll Results*, June 2014; http://online.wsj.com/articles/the-wall-street-journalnbc-news-poll-1378786510?tesla=y. Accessed December 1, 2014.

40. "American Lung Association Cleaner Gasoline and Vehicle Standards Survey Findings," *Greenberg Quinlan Rosner Research*, January 2013; http://www.lung.org/healthy-air/outdoor/resources/cleaner-gasoline-and-vehicles-survey-powerpoint.pdf. Accessed December 1, 2014.

41. "New Poll Shows the Public Wants EPA to Do More to Reduce Air Pollution," *American Lung Association*, March 2012; http://www.lung.org/press-room/press-releases/new-poll-shows-the-public.html. Accessed December 1, 2014.

42. Anthony Leiserowitz, Edward Maibach, Connie Roser-Renouf, and Nicholas Smith, "Climate Change in the American Mind: Americans' Global Warming Beliefs and Attitudes in May 2011," *Yale University and George Mason University*, April/May 2011, 14–15; http://environment.yale.edu/climate-communication/files/Americans-Global-Warming-Beliefs-and-Attitudes-in-May-2011.pdf. Accessed December 1, 2014.

43. George C. Edwards, Martin P. Wattenberg, and Robert L. Lineberry, *Government in America: People, Politics, and* Policy (Upper Saddle River, NJ: Pearson Custom Publishing, 2013).

44. "2014 Federal Employee Viewpoint Survey Results for the U.S. Environmental Protection Agency," *The Office of Personnel Management*, April-June 2013; http://www.epa.gov/ohr/2014-evs-web-summary-new.pdf/. Accessed December 1, 2014; "2014 Federal Employee Viewpoint Survey Results: Governmentwide Management Report," *The Office of Personnel Management*, 2014, Appendix H; http://www.fedview. opm. gov/2014files/2014_Government wide_Management_Report.PDF. Accessed December 2, 2014; "2013 Federal Employee Viewpoint Survey Results for the U.S. Environmental Protection Agency," *The Office of Personnel Management*, April-June 2012; http://www.epa. gov/ohr/2013-evs-web-summary-new.pdf/. Accessed November 18, 2014.

Chapter 7. The Environmental Court

We particularly would like to thank Alena Smith of Brigham Young University, our research assistant, for assisting in collecting material for this chapter.

1. "Supreme Court Clerk Hiring Watch: The Official List for October Term 2013," *Above the Law*; http://abovethelaw.com/2013/07/supreme-court-clerk-hiring-watch-the-official-list-for-october-term-2013/. Accessed January 23, 2014.

2. Congressional Research Service Report RL30064, *Congressional Salaries and Allowances*, by Ida A. Brudnick (January 4, 2012), 4. Text in library clerk house.gov/reference-files/112_20120104_Salary.pdf.

3. Congressional Research Service Report RL30807, *Congressional Member Office Operations*, by John S. Pontius, 3. Updated December 21, 2004. Text in opencrs.com/document/RL30807/.

4. Robert V. Percival, "Environmental Law in the Supreme Court: Highlights from the Marshall Papers," *Environmental Law Reporter* 23 (October 1993): 10607.

5. Examples of such critics include Donald Horowitz, *The Courts and Social Policy* (Washington, DC: Brookings, 1977); and R. Shep Melnick, *Regulation and the Courts*

(Washington, DC: Brookings, 1983). As well, Lettie McSpadden, in a chapter of Vig and Kraft's book makes a similar argument. See "Environmental Policy in the Courts," *Environmental Policy: New Directions for the Twenty-First Century*, ed. in Norman J. Vig and Michael E. Kraft (Washington, DC: CQ Press, 2000), 145.

6. Rachel Carson, *Silent Spring* (Boston: Houghton Mifflin, 1962).

7. See a comparison of the environmental issues that became a focus for Franklin Roosevelt and Bill Clinton in Byron W. Daynes, "Two Democrats, One Environment: First-Term Efforts of Franklin Roosevelt and Bill Clinton to Shape the Environment," in Byron W. Daynes, William D. Pederson, and Michael P. Riccards, *The New Deal and Public Policy* (New York: St. Martin's Press, 1998), esp. 113–17.

8. Alexis de Tocqueville, *Democracy in America*, vol. 1, ed. J. P. Mayer, trans. George Lawrence (Garden City, NY: Doubleday Anchor, 1969), 270.

9. Lettie M. Wenner, *The Environmental Decade in Court* (Bloomington: Indiana University Press, 1982), 2–3.

10. McSpadden, "Environmental Policy in the Courts," 145.

11. Kenneth M. Holland, "The Role of the Courts in the Making and Administration of Environmental Policy in the United States," in Kenneth M. Holland, F. L. Morton and Brian Galligan, *Federalism and the Environment* (Westport, CT: Greenwood, 1996), 164.

12. 420 US 35 (1975).

13. 458 US 941 (1982).

14. 427 US 390 (1976).

15. 503 US 91 (1992).

16. 540 US 461 (2004).

17. 420 US 136 (1975).

18. 434 US 275 (1978).

19. 549 US 497 (2007).

20. 564 US ____ (2011).

21. 572 U.S. _____ (2014).

22. 573 U.S. ____ 2014.

23. Justice Scalia majority opinion in *Utility Air Regulatory Group v. EPA*, 573 U.S. ____ 2014.

24. 454 US 139 (1981).

25. 456 US 305 (1982).

26. 555 US 7 (2008).

27. 478 US 221 (1986).

28. 497 US 871 (1990).

29. 464 US 312 (1984).

30. 479 US 470 (1987).

31. 554 US 471 (2008).

32. 563 US ____ (2011).

33. 422 US 655 (1973).

34. 474 US 121 (1985).

35. 575 US 99 (2002).

36. 501 US 597 (1981).

37. 295 US 1 (1935).

38. 315 US 681 (1942).

39. 391 US 392 (1968).
40. 480 US 470 (1987).
41. 545 US 75 (2005).
42. 569 US ____ (2013).
43. 416 US 861 (1974).
44. 437 US 617 (1978).
45. 505 US 144 (1992).
46. 405 US 727 (1972).
47. 427 US 246 (1976).
48. 411 US 325 (1973).
49. 448 US 242 (1980).
50. 474 US 494 (1986).
51. 446 US 657 (1980).
52. 491 US 1 (1989).
53. 444 US 164 (1979).
54. Percival, "Environmental Law in the Supreme Court," 10613. Percival notes, "These statistics were derived by analyzing cases listed as environmental cases in *U.S. Law Week* for the period from 1970 to 1991."
55. William Funk, "Justice Breyer and Environmental Law," *Administrative Law Journal* (American University) 8 (1995): 735, 741–43. Reference in Daniel A. Farber, "Is the Supreme Court Irrelevant? Reflections on the Judicial Role in Environmental Law," *Minnesota Law Review* 8 (February 1997): 565.
56. Percival, "Environmental Law in the Supreme Court," 10625.
57. 497 U. S. 871 (1990).
58. The following two decision illustrate this: *Lujan v. National Wildlife Federation,* 110 S. Ct. 3177 (1990); and *California v. Federal Energy Regulatory Commission* 110 S. Ct. 2024 (1990). See also O'Leary, "Environmental Policy in the Courts," 125.
59. Pub. L. 91-190. See also Ray Clark and Larry Canter, ed., *Environmental Policy and NEPA* (Boca Raton: St. Lucie Press, 1997), 17.
60. Pub.L. 101-549.
61. Pub. L. 100-4.
62. Pub. L. 92-531.
63. Pub. L. 92-574.
64. Pub. L. 93-205, and more recent amendments.
65. Pub. L. 92-516, and more recent amendments.
66. Pub. L. 92-583, and more recent amendments.
67. Pub. L. 94-469.
68. Tocqueville, *Democracy in America,* vol. 1, 150.
69. Thomas R. Marshall, *Public Opinion and the Supreme Court* (Boston: Unwin Hyman, 1989), 71.
70. Robert Weissberg, *Public Opinion and Popular Government* (Englewood Cliffs, NJ: Prentice-Hall, 1976), 121–26.
71. Marshall, *Public Opinion and the Supreme Court,* 78.
72. *Gallup Poll,* July 19–22, 2012; http://www.gallup.com/poll/156347/americans-next-president-prioritize-jobs-corruption.aspx. Accessed January 30, 2014.
73. Dunlap, "Public Opinion and Environmental Policy," 94.

74. Stephen L. Wasby, *The Supreme Court in the Federal Judicial System* (Chicago: Nelson-Hall Publisher, 1994), 3.

75. Richard E. Levy and Robert L. Glicksman, "Judicial Activism and Restraint in the Supreme Court's Environmental Law Decisions," *Vanderbilt Law Review* 42 (March 1989), 346.

76. 437 U. S. 153 (1978).

77. See Albert Gore, *Earth in the Balance: Ecology and the Human Spirit* (Boston: Houghton Mifflin, 1992); as well as Gore's Web statement: www.algore2000.com/agenda/issue_environ.html. Accessed May 25, 2000. Also see: www.algore2000...ng_to_protect_our_environment.html. Accessed May 25, 2000.

Chapter 8. The Global Environment

1. Jonathan P. West and Glen Sussman, "Implementation of Environmental Policy: The Chief Executive," in *The Environmental Presidency*, ed. Dennis L. Soden (Albany: State University of New York Press, 1999), 102–104.

2. Sheila Jasanoff, *The Fifth Branch: Science Advisors as Policymakers* (Cambridge: Harvard University Press, 1990), 250.

3. Lynton Keith Caldwell, *Between Two Worlds: Science, the Environmental Movement, and Policy Choice* (New York: Cambridge University Press, 1990), 19–20.

4. Andrew L. Dessler and Edward A. Parson, *The Science and Politics of Global Climate Change* (Cambridge: Cambridge University Press, 2006), 39.

5. James Rosenau, "Environmental Challenges in a Global Context," In Sheldon Kamieniecki, *Environmental Politics in the International Arena* (Albany: State University of New York Press, 1993), 257.

6. Ibid.

7. Lynton Keith Caldwell, *International Environmental Policy* (Durham: Duke University Press, 1996), 5–10.

8. Chad J. McGuire, *Adapting to Sea Level Rise in the Coastal Zone* (Boca Raton: CRC Press, 2013), 42.

9. Orrin Pilkey and Rob Young, *The Rising Sea* (Washington: Island Press, 2009), 34.

10. National Oceanic and Atmospheric Administration, "Arctic Report Card: Region Continues to Warm at Unprecedented Rate," October 21, 2010; wwwnoaanews.noaa.gov/stories2010/20101021_arcticreportcard.html.

11. David A. Gabel, "How Rising Sea Levels Will Affect the U.S. Coastline," *Environmental News Network*, February 18, 2011; www.enn.com.

12. U.S. National Research Council and U.S. National Oceanic and Atmospheric Administration, *America's Climate Choices: Panel on Adapting to the Impacts of Climate Change* (Washington, DC: The National Academies Press, 2010.

13. Oran R. Young, "Rights, Rules, and Resources in World Affairs," in *Global Governance: Drawing Insights from the Environmental Experience*, ed. Organ R. Young, (Cambridge:The MIT Press, 1997), 7–9.

14. James Connelly and Graham Smith, *Politics and the Environment* (New York: Routledge, 1999), 198.

15. Young, *Global Governance*, 98; David G. Victor, Abram Chayes, and Eugene B. Sklonikoff, "Pragmatic Approaches to Regime Building for Complex International Problems," in *Global Accord: Environmental Challenges and International Responses*, ed. Nazli Choucri (Cambridge: The MIT Press, 1993), 463; United National Environment Program, "UNEP Achievements," August 2000; www.unep.org/Documents/Default.asp?DocumentID=43&ArticleID=250. Accessed August 16, 2000; UNEP, "The State of the Planet's Biodiversity," May 22, 2010; www.unep.org/wed/2010/english/biodiversity.asp. Accessed November 27, 2014.

16. James Gustave Speth and Peter M. Haas, *Global Environmental Governance* (Washington, DC: Island Press, 2006), 111.

17. Kurt Tudyka, "Regulations Problems of a General European Environmental Policy," in *Environmental Policy in Europe*, ed. Markus Jachtenfuchs and Michael Strubel (Baden-Baden: Nomos Verlagsgesellschaft, 1992), 234; United Nations Environment Program, *Global Governance Outlook* (New York: Oxford University Press, 1997), 75.

18. Gary Bryner, *From Promises to Performance* (New York: W. W. Norton, 1997), 15.

19. Lamar Hampel, *Environmental Governance* (Washington, DC: Island Press, 1996), 30.

20. Stanley P. Johnson, *The Earth Summit: The United Nations Conference on Environment and Development* (Boston: Graham and Trotman/Martinus Nijhoff, 1993), 62.

21. Glen Sussman and Byron W. Daynes, *U.S. Politics and Climate Change* (Boulder: Lynne Rienner, 2013), 37–38.

22. Paul G. Harris, "International Environmental Affairs," in *The Environment, International Relations, and U.S. Foreign Policy*, ed. Paul G. Harris (Washington, DC: Georgetown University Press, 2001), 18.

23. Theodore C. Sorenson, *Kennedy* (New York: Harper and Row, 1965), 621.

24. Charles H. Southwick, *Global Ecology in Human Perspective* (New York and Oxford; Oxford University Press, 1996), 318.

25. Dan Lamothe, "Climate Change Threatens National Security, Pentagon Says," *The Washington Post*, October 13, 2014; www.washingtonpost.com/news/checkpoint/wp/2014/10/13/climate-change-threatens-national-security, Pentagon-says. Accessed November 16, 2014.

26. Chuck Hagel, "Department of Defense: 2014 Climate Change Adaptation Roadmap." Office of the Deputy Under Secretary Secretary of Defense for Installations and Environment, Alexandria, Virginia, 2014.

27. *The Global 2000 Report to the President: Entering the Twenty-First Century* (New York: Penguin Books, 1982), preface.

28. Ibid, 1.

29. Vaclav Smil, "Our Changing Environment," *Current History* 88 (January 1989): 48.

30. John Noble Wilford, "Ages Old Icecap at North Pole Is Now Liquid, Scientists Find," *New York Times*, August 19, 2000, A1, A13.

31. Byron W. Daynes and Glen Sussman, *White House Politics and the Environment* (College Station: Texas A&M University Press, 2010), 193.

32. Mark Hertsgaard, *Earth Odyssey: Around the World in Search of OurEnvironmental Future* (New York: Broadway Books, 1998), 335.

33. Thomas L. Friedman, *Hot, Flat, and Crowded* (New York: Farrar, Straus and Giroux, 2008), 5.

34. Jasanoff, *The Fifth Branch*, 250.

35. Glen Sussman, "The USA and Global Environmental Policy: Domestic Constraints on Effective Leadership," *International Political Science Review* 25 (October 2004): 365.

Chapter 9. Conclusion

1. Eric H. Cline, "Climate Change Doomed the Ancients," *New York Times Op-Ed*, May 28, 2014, A21.

2. Jedediah Purdy, "Shades of Green," *The American Prospect*, January 3, 2000, 6.

3. Marian R. Chertow and Daniel C. Esty suggest that the polls seem to indicate about 80 percent of Americans think of themselves as environmentalists. See Marian R. Chertow and Daniel C. Esty, "Environmental Policy: The Next Generation," *Issues in Science and Technology Online*, Fall 1997; http//bob.nap.edu/issues/14.1/esty.htm. Accessed September 13, 2000.

4. Purdy, "Shades of Green," 6.

5. Ibid.

6. Ibid.

7. See Gus diZerega, "Stewart Brand II: Romantic Environmentalism"; https://www.google.com/search?q=http%3A%2F%2Fwww.beliefnet.com%2Fcolumnists%2Fpagansblog%2F2006%2F12%2Fstewart-brand-ii-romantic-environmentalism.html&oq=http%3A%2F%2Fwww.beliefnet.com%2Fcolumnists%2Fpagansblog%2F2006%2F12%2Fstewart-brand-ii-romantic-environmentalism.html&aqs=chrome..69i58j69i57.82552j0j8&sourceid=chrome&es_sm=93&ie=UTF-8. Accessed November 7, 2014.

8. Executive Office of the President, "The President's Climate Action Plan," (Washington, DC: The White House, June 2013), 4.

9. Al Gore, *Earth in the Balance: Ecology and the Human Spirit* (Boston: Houghton Mifflin, 1992).

10. Douglas Jehl, "West Takes a Stand Against Clinton Plan," *New York Times* on the Web, September 14, 2000. Accessed September 14, 2000.

11. Robert Keiter, director of the Wallace Stegner Center for Land, Resources, and the Environment, suggested this at a conference on "Learning from the Monument: What Does the Grand Staircase-Escalante Mean for Land Protection in the West," at the University of Utah Law School, Salt Lake City, Utah, September 15, 2000.

12. Jehl, "West Takes a Stand."

13. Juliet Eilperin, "National Ocean Policy Sparks Partisan Fight." *Washington Post*, October 28, 2012; http://www.washigntonpost.com/national/heath-science/national-ocean-policy-sparks-partisan-fight/2012/10/28/af73e464-17a7-11e2-a55c-39408fbe-6a4b_story.html. Accessed November 9, 2014.

14. Barack Obama, Second Inaugural Address, January 2013 (Washington, DC: The White House, 2013), 4.

15. "The President's Climate Action Plan."

16. Mark Lander, "U.S. and China Reach Agreement on Climate In Step to Global Pact," *New York Times,* November 12, 2014, A1.

17. Ibid., A11.

18. Christopher Marquis, "Bush Administration Rolls Back Clinton Rules for Wetlands," *New York Times,* January 15, 2002, 16A.

19. "Bush's First 100 Days Set a Terrible Precedent on Environmental Policy," PBS Newshour. 2001. "Bush and the Environment." *Public Broadcasting System,* March 29; http://www.pbs.org/newshour/bb/environment-jan-june01-bushenv_3-29/.

20. Katharine Q. Seelye, "Bush Team Is Reversing Environmental Policies." *New York Times,* November 18, 2001; http://www.nytimes.com/2001/11/18/politics/18ENV8Lhtml.

21. Douglas Jehl, "On Rules for Environment, Bush Sees a Balance, Critics a Threat" *The New York Times,* February 23, 2003, A1.

22. Kraft, *Environmental Policy and Politics,* 62–63.

23. Tom Hamburger, "Sen. Inhofe, Denier of Human Role in Climate Change, Likely to Lead Environment Committee," *Washington Post,* November 5, 2014; http://www.washingtonpost.com/politics/inhofe-an-epa-foe-likely-to-lead-senate-environment-committee/2014/11/05/d0b4221e-64f4-11e4-836c4f26eb67_story.html. Accessed November 19, 2014.

24. 549 U.S. 497 (2007).

25. 564 U.S. _____ (2011).

26. 420 U.S. 35 (1975).

27. 555 U.S. 7 (2008).

28. 479 U.S. 481 (1987).

29. 295 U.S. 1 (1935).

30. Nicholas Lutsey and Daniel Sperling, "America's Bottom-up Climate Change Mitigation Policy," *Energy Policy* 36, no. 2 (February 2008): 673–85.

31. Morgan Jacobsen, "Poor Air Quality Puts Physical, Economic Health at Risk, Salt Lake City Mayor Says," *Deseret News,* January 13, 2014; http://www.deseretnews.com/article/865593783/Poor-air-quality-puts-physical-economic-health-at-risk-Salt-Lake-City-mayor-says.html?pg=all. Accessed November 22, 2014.

32. Frank Newport, "Smaller Majorities in U.S. Favor Government Pollution Controls." Gallup, June 4, 2014; http://www.gallup.com/poll/170885/smaller-majorities-favor-gov-pollution-controls.aspx. Accessed November 22, 2014.

33. Swift, "Americans Again Pick Environment Over Economic Growth."

34. Rebecca Riffkin, "Climate Change Not a Top Worry in U.S." Gallup, March 12, 2014; http://www.gallup.com/poll/167843/climate-change-not-top-worry.aspx. Accessed, November 22, 2014.

35. See Riley E. Dunlap, "Trends in Public Opinion toward Environmental Issues: 1965–1990," in *American Environmentalism: The U. S. Environmental Movement, 1970–1990.,* ed. Riley E. Dunlap and Angela G. Mertig (New York: Taylor and Francis, 1992), 93.

36. *PollingReport.com.* "Problems and Priorities." *CNN/ORC Poll,* January 14–15, 2013; http://www.pollingreport.com/priority.htm. Accessed November 24, 2014.

37. *PollingReport.com.* "Problems and Priorities." *Quinnipiac University Poll,* January 30-February 4, 2013; http://www.pollingreport.com/priority.htm. Accessed November 24, 2014.

38. Dunlap, "Public Opinion and Environmental Policy," 94.

39. Thomas R. Marshall, *Public Opinion and the Supreme Court* (Boston: Unwin Hyman, 1989), 78.

40. Samuel P. Hays, *Explorations in Environmental History* (Pittsburgh: University of Pittsburgh Press, 1998), 384.

41. An example of a controversial advertisement used by the People for the Ethical Treatment of Animals recently was their "Got Beer?" advertisement which was directed to college students that cautioned students about drinking of milk, suggesting that it might be responsible for breast cancer and prostate cancer. The ad indicated that it might be safer for students to drink beer than to continue to drink milk. It had a very short run due to the controversy it stirred up. See "Opinion: PETA's Controversial Ad Campaign was Irresponsible," *NewsNet@BYU,* March 20, 2000; http://newsnet.byu.edu/print_story. efm?number=8527&year=current. Accessed September 13, 2000.

42. Steve Brodner, 2014. "The Tar Sands Pipeline of Politics," *The Nation,* November 10, 2014, 21. Brodner's source was *#Disrupt Denial Report* by *Forecast the Facts Action* and *Sum of Us* using reports filed to the FEC and compiled by *OpenSecrets.org.*

43. Allan Freedman, "Prospects in Senate Brighten for Rewrite of Species Law," *Congressional Quarterly Weekly Report* 55, no. 20 (1997): 1125.

44. Swift, "Americans Again Pick Environment Over Economic Growth."

45. See chapter 4.

46. See Allan Freedman, "Tensions Mount within House GOP," *Congressional Quarterly Weekly Report* 55, no. 20 (1997): 1126.

47. Joe Manchin III, League of Conservation Voters Scorecard, 2013; http://score-card.lcv.org/moc/joe-manchin-iii. Accessed December 8, 2014.

48. Allan Freedman, "Tensions Mount within House GOP," 1126.

49. Both Robert Keiter, director of the Wallace Stegner Center for Land, Resources and the Environment, and David Williams, a Wallace Stegner Fellow of the center suggested this at a conference on "Learning from the Monument: What Does the Grand Staircase-Escalante Mean for Land Protection in the West," at the University of Utah Law School, Salt Lake City, Utah, September 15, 2000.

50. Marian R. Chertow and Daniel C. Esty suggest that the polls they have seen indicate about 80 percent of Americans think of themselves as environmentalists. See Marian R. Chertow and Daniel C. Esty, "Environmental Policy: The Next Generation," *Issues in Science and Technology Online,* Fall 1997; http//bob.nap.edu/issues/14.1/esty. htm. Accessed September 13, 2000.

51. Ibid.

52. Lynton Keith Caldwell, *International Environmental Policy: From the Twentieth to the Twenty-First Century* (Durham: Duke University Press, 1996), 5–10.

53. See *The Global 2000 Report to the President: Entering the Twenty-First Century* (New York: Penguin Books, 1982).

Index

acid rain, 16, 84–85, 119, 120, 174–75, 207

Adamo Wrecking Co. v. U.S. (1978), 152–53

adjudication, 135

Administration Procedure Act (APA), 130–31, 135

Africa, 184, 189, 201

Air Pollution Variance Board v. Western Alfalfa Corporation (1974), 156

Alaska, 13, 155, 215

Alaska Department of Environmental Conservation v. EPA (2004), 152

Alaska National Interest Lands Conservation Act (1980), 13, 63, 108

Alaska v. United States (2005), 155

Albright, Madeleine, 180

Alito, Samuel, 214

alternative energy, 52–54, 55, 60

American Electric Power v. Connecticut (2011), 153, 167, 214

American Farm Bureau (AFB), 65, 67, 114

American Petroleum Institute, 6

American Trucking Assoc. v. City of Los Angeles, California (2013), 155

Anderson, Rocky, 216

Andrews, Richard N. L., 92

Andrus, Cecil, 108

Andrus v. Shell Oil Company (1980), 157

Antarctica, 197

Antiquities Act (1906), 14, 208

ARCO, 31

Arctic, 197

Arctic National Wildlife Refuge (ANWR), 14–15, 211

Arizona, 216

Arkansas, 152

Arkansas v. Oklahoma (1992), 152

Arnold, R. Douglas, 92

Askew v. American Waterways Operators (1973), 157

Australia, 119, 189

Babbitt, Bruce, 66, 114, 208

Ban Ki-Moon, 180

Bangladesh, 176

Bast, Joseph, 194

Beck, Glenn, 56

biodiversity, 17, 50, 59, 63, 177, 199–201

Blackmun, Harry, 160, 214, 222

Boehlert, Sherwood, 221

Boots, Mike, 115

Border Ranch Partnership v. U.S. Army Corp of Engineers (2002), 155

Bosso, Christopher, 58

BP oil spill, 82, 85–86, 90, 128

Brazil, 200

Brennan, William J., Jr., 214, 222

Breyer, Stephen G., 158, 214

Brown, Jerry, 28

Brown, Scott, 86
Browner, Carol, 114, 115
Bureau of Indian Affairs, 126
Bureau of Land Management (BLM),
    126, 154
Bureau of Reclamation, 126
Burford, Anne, 113, 121, 210
Bush, George H. W.
    and acid rain, 16, 84–85, 119
    and clean air, 14, 110–11, 208
    and international agreements, 6–7,
        119, 180
    and legacy, 14, 15, 103, 111, 163
Bush, George W.
    as chief executive, 4, 15, 114,
        115–16
    and climate change, 16, 61, 166–67
    and Congress, 109, 111
    and international agreements, 16,
        166–67, 180, 197–98
    legacy of, 14–15, 61, 211–12, 216
    policies of, 61, 66, 211–12
    and public opinion, 57, 68, 218
Byrd, Robert, 35–36, 221

Caldwell, Lynton, 172, 199
California, 27, 28–29, 31, 86, 154,
    215, 216, 221
California Air Resources Board
    (CARB), 27, 28–29
California Desert Protection Act
    (1994), 14
Cameron, Charles M., 110
Canada
    and acid rain, 16, 120
    and emission reductions, 189, 192
    and international cooperation, 12,
        119, 193, 200
    and Kyoto Protocol, 180, 188, 194
    and public opinion, 173
Cannon, Jonathan Z., 89
cap-and-trade, 29, 84–86, 131
Capito, Shelley M., 221

Carbon Limits and Energy for
    American Renewal (CLEAR) Act,
    81
Carter, Jimmy, 108, 110, 112, 196
Carson, Rachel, 12, 23, 63, 150
Cater, Douglas, 135
Cato Institute, 6, 57
Center for Biological Diversity, 67
Cheney, Dick, 114
Chertow, Marian, 222
Chesapeake Bay, 15
Chile, 173
China, 119, 131, 184, 210
Chu, Steven, 127–28
CITES, 16, 200
City of Philadelphia v. New Jersey
    (1978), 156
Civilian Conservation Corps (CCC),
    11–12, 99–100, 107, 208
Clark, William, 126
Clean Air Act
    creation of, 3, 13, 107, 161, 212
    and enforcement, 36, 105, 131,
        153, 167
    weakening of, 15, 109, 211
Clean Air Act Amendments (1990),
    14, 16, 25–26, 111, 161, 208,
    212
clean energy. *See* alternative energy
Clean Water Act, 3, 13, 15, 34, 36,
    109
Clean Water Act Amendments, 14, 108
Clear Skies Initiative, 15, 211
Climate Action Plan, 209
climate change
    and executive agencies, 127–28, 132,
        138, 167, 171
    and George W. Bush, 15, 166–67
    global, 132, 177, 178–80, 181–82,
        205, 223
    and interest groups, 59, 63
    and Obama, 57, 104–6, 131–32,
        206, 209–10, 222

and public opinion, 7, 17, 50–52, 183–84, 217
and sea level rise, 39–42, 175–76
and Supreme Court, 166–68
Cline, Eric H., 205
Clinton, William J.
  as chief executive, 112–13, 114, 115, 159, 208
  and Congress, 61, 108, 109, 111, 220
  as environmentalist, 14, 15, 103, 150, 206, 208–9
  and international agreements, 16, 166
coal energy, 53, 104, 127, 132, 221
coal industry, 33–37, 56, 106, 133, 155
Coastal Zone Management Act (1972), 161
Collins, Susan, 81–83, 84
Congress
  behavioral dynamics in, 84–85, 87–92
  committees, 3, 72, 73–74, 77–79, 212–13
  decision processes in, 80–81, 90
  and environmental legislation, 3, 12, 13–14, 207, 212
  and executive agencies, 80, 124, 140, 213
  operational characteristics of, 78–79
  partisanship in, 76, 77, 84–85, 104, 220–21, 222
  and the president, 61, 107–11, 151–52, 213
  process of, 2, 18, 71–73
  and science, 90–91, 93
  structural factors of, 73–78
Connelly, Joel, 108
conservation, 8, 9, 10–11, 17, 206, 208
Costa Rica, 119
Council on Environmental Quality (CEQ), 125, 130, 171, 196

Davis, David, 72, 84
Daynes, Byron, 50–51, 101, 114, 184
DeConcini, Christina, 40
Defenders of Wildlife, 59, 65, 219
Delaware, 154
delegated programs, 24
Democratic Party
  in Congress, 15, 84, 104, 108, 220–21
  and interest groups, 62
  in other offices, 15, 61, 222
  and public opinion, 48–50, 51–52, 53
Denmark, 173
Department of Agriculture, 133
Department of Defense, 181, 182
Department of Energy (DOE), 125, 126–27, 133
Department of the Interior (DOI)
  and environmental policy, 15, 19, 125, 126, 154, 171
  and politicization, 4, 114
Department of State, 196
Dessler, Andrew L., 172, 184
de Tocqueville, Alexis, 150, 162
Division of Grazing, 100
Douglas, William O., 214, 222
Dowie, Mark, 114
Downes, William, 65
Downs, Anthony, 47–48
Dunlap, Riley, 163, 218
Durant, Robert, 4, 138

E. I. DuPont de Nemours & Co. v. Train (1976), 152
Earth Day, 13, 61
Earth First!, 5, 63, 64
Earth Summit (1992), 6–7, 16, 119, 179, 200
Edwards, George C., 140–41
Ehrlich, Paul, 23
Ehrlichman, John, 103
Eisenhower, Dwight D., 12, 102

emissions, 29, 131–33, 152–53, 167
  reductions in, 180, 185, 187–91,
    192, 209, 210
Endangered Species Act
  creation of, 3, 13
  enforcement of, 66, 161, 163–65
  weakening of, 68, 88, 221
Endangered Species Reform Coalition,
  68
energy subsidies, 55
enforcement, 26, 54–55
Environment America (EA), 84
Environmental Defense Fund, 59, 63,
  219
Environmental Protection Agency (EPA)
  and Congress, 80, 140, 207, 213
  creation of, 13, 23, 61, 108,
    137–38, 208
  and enforcement, 26, 31, 32, 125,
    131–33, 167
  and environmental policy, 19, 60,
    137–38, 207
  and the president, 4, 106, 112, 113,
    131, 198
  and public opinion, 138–41
  and stakeholders, 138–40, 141–44,
    145
  and Supreme Court, 131, 152–53,
    167, 213–14
environmentalism, 205–7
EPA v. EME Homer City Generation
  (2014), 153, 167
Esty, Daniel, 222
Euchner, Charles, 58
European Union (EU), 177–78, 189,
  192–93, 223
executive agencies, 3–4, 19, 123–25,
  130–31, 135–36, 152–53, 207.
  See also specific agencies
executive orders, 115–16
Exxon, 60, 90

Federal Emergency Management Agency
  (FEMA), 82, 112, 128, 171

Federal Employee Viewpoint Survey
  (EVS), 141–44
Federal Pesticide Control Act (1972),
  161
Federal Regulatory Commission, 133
federalism, 2, 18, 21
Feinstein, Dianne, 14
Fingar, Thomas, 180
Fiorino, Daniel J., 78
Fish and Wildlife Service (FWS), 15,
  65, 67, 126
Flood Control Act (1936), 12
Florida, 31, 39–42
fracking, 53, 109
Friedman, Thomas L, 198–99
Friends of the Earth, 56, 59, 60, 218,
  219
Friends of the Earth v. Laidlaw
  Environmental Services (2000),
  159, 160

Germany, 173
Ginsburg, Ruth Bader, 158, 159, 160,
  214
Glicksman, Robert L., 163
global warming. See climate change
Goldwater, Barry, 221
Gonzalez, George, 132–33
Gore, Al, 112, 114, 207, 208
Gorsuch, Anne, 4
Grand Staircase—Escalante National
  Monument (UT), 14, 63
Greece, 188
green business, 60
Greenpeace, 5, 63–64, 219
greenwashing, 68

Hagel, Chuck, 181–82
Hair, Jay, 61
Hamilton, Alexander, 124
Hannity, Sean, 56
Harkin, Tom, 221
Harper, Stephen, 119
Harrison, Kathryn, 183

Harrison, Paul, 180–81, 199
Hathaway, Stanley, 126
Hawaii, 216
Hays, Samuel, 9, 218
Healthy Forest Initiative, 15, 211–12
Heartland Institute, 6
Hempel, Lamont, 8
Hertsgaard, Mark, 198
Hickel, Walter, 126
Holland, Kenneth, 151
Holmstead, Jeffrey, 114
Huntsman, Jon, 216
Hurricane Sandy, 105, 128

Iceland, 189
Ickes, Harold Le Clair, 100, 113
Idaho, 215
Indiana, 30
Ingleheart, Ronald, 9
Inhofe, James M., 213, 219
interest groups
  environmental, 5–6, 55–56, 58–64, 211, 218–19
  industry, 6, 67–68, 219–20
Intergovernmental Panel on Climate Change (IPCC), 175, 177, 184, 197
international agreements, 12, 16, 131, 132, 179–80, 197–98, 200
  See also Earth Summit; Kyoto Protocol; Montreal Protocol
International Convention on Trade in Endangered Species of Wild Fauna and Flora. See CITES
International Paper v. Ouelette (1987), 154, 214
iron triangles, 135–36
Israel, 188
issue networks, 136

Jackson, Lisa, 4, 28, 114–15
Jacobsen, Morgan, 216–17
Japan, 119, 153–54, 173

Japan Whaling Assn. v American Cetacean Society (1986), 154
Jasanoff, Sheila, 171–72, 199
Jenkins, David, 82
Jewell, Sally, 115
Johnson, Lyndon B., 107, 113
Joseph, Peniel E., 116

Kagan, Elena, 214
Kaiser Aetna v. U.S. (1979), 157
Kelso, Mark, 10, 77, 111
Kennedy, Anthony M., 158, 214, 215, 222
Kennedy, John F., 13, 16, 63, 103, 181
Kennedy, Robert F., Jr., 56–57
Kenya, 200
Kerry, John, 104, 115, 222
Key West, FL, 40
Keystone Bituminous Coal v. De Benedictis (1987), 155
Keystone XL oil pipeline, 56, 82, 109, 219
Kirk, Mark, 84
Klein, Naomi, 185
Kovacic, William E., 161
Kraft, Michael, 75, 76, 121, 212
Kyoto Protocol
  global involvement in, 185–86, 187, 189, 192–93, 194, 195
  and presidents, 119, 166, 179–80, 211
  and states, 216

Landler, Mark, 210
Lavelle, Rita, 4
League of Conservation Voters (LCV), 82, 84, 108–9, 219, 221
Leiserowitz, Anthony, 183
Lester, James P., 5
Levine, Philip, 40
Levy, Richard E., 163
Lieberman, Joseph, 82
Limbaugh, Rush, 56, 166

Limited Nuclear Test Ban Treaty
(1963), 16, 179, 181
Los Angeles, CA, 155
Lujan v. Defenders of Wildlife (1992),
160
Lujan v. National Wildlife Federation
(1990), 154, 158, 159
Lutsey, Nicholas, 216
Lyons, Michael, 88, 89–90

Madison, James, 124
Maine, 27, 81, 215
Maldives, 176
Manchin, Joe, III, 221
mandated programs, 24, 25–27
Marbury v. Madison (1803), 4
Marine Mammal Protection Act
(1972), 3
Marine Preservation Association, 6, 68
Marine Protection, Research and
Sanctuaries Act (1972), 161
Marshall, Thomas R., 162–63, 218
Marshall, Thurgood, 214, 222
Massachusetts, 31, 216
Massachusetts v. EPA (2007), 131, 153,
167, 214
Massaro, Tony, 82
Mayhew, David R., 88
McCain, John, 67
McCarthy, Gina, 115
McKibben, Bill, 57
Mexico, 12, 119, 173, 188, 193, 200
Miami, FL, 39, 40, 176
Miami Beach, FL, 40
*Midatlantic National Bank v. New
Jersey Department of Environmental
Protection* (1986), 157
Miller, G. Tyler, 200
Minnesota, 27, 31, 33, 216
Missouri, 26
Moniz, Ernest, 115
Montana, 154, 215
Montana v. Wyoming and North Dakota
(2011), 154

Montreal Protocol on Ozone Depletion
(1987), 16, 118, 119
mountaintop removal, 33–38
Muir, John, 8–9, 10, 11, 206, 218

Nathan, Richard, 113
National Association of Manufacturers,
6
National Audubon Society, 59, 60, 65,
219
National Environmental Performance
Partnership System (NEPPS), 32
National Environmental Policy Act
(NEPA) of 1970
creation of, 3, 13, 61, 107–8, 212
and implementation, 130, 161, 215
weakening of, 109
national forests, 156, 208, 212
national monuments, 14, 63, 208–9
National Ocean Council, 209
National Oceanic and Atmospheric
Administration (NOAA), 175,
176
National Park Service, 126
national parks, 99, 100, 112, 208, 211
National Research Council, 176
National Resources Board, 100
National Wetlands Coalition, 6, 68
National Wildlife Federation
creation of, 10, 59–60, 63, 219
and environmental issues, 5, 61, 63,
65, 66, 154
natural gas, 53
Natural Resources Defense Council,
56, 59, 219
Nelson, Bill, 39
Nevada, 215
New Jersey, 31, 154, 156, 216
New Jersey v. Delaware (2008), 154
New Orleans, LA, 176
New York, 26, 154, 156, 214
New York v. U.S. (1992), 156
New Zealand, 119
Nichols, Mary, 28–29

Nixon, Richard M.
    appointments by, 222
    and Congress, 110, 111, 121
    as environmentalist, 102–3, 107–8,
        121, 208, 221
    and EPA, 23, 61, 108, 112, 137,
        208
    and legislation, 13, 61, 107–8, 200
Noise Control Act (1972), 161
North American Free Trade Act
    (NAFTA), 16
North Carolina, 216
North Dakota, 154
Norton, Gale, 4
Norway, 119, 189
nuclear energy, 53–54, 55, 127
Nuclear Regulatory Commission
    (NRC), 125, 129–30
Nuclear Waste Policy Amendments
    (1987), 108

Obama, Barack
    and cap-and-trade, 85–86, 131
    as chief executive, 4, 15, 114–15,
        116, 127, 128, 198
    and climate change, 40, 57, 103–6,
        131–32, 206, 209–10, 222
    and Congress, 16, 37, 92, 108–9,
        111, 213
    and the EPA, 106, 112, 114–15,
        131–32, 198
    international agreements of, 16, 119,
        131, 132, 180
    and public opinion, 56, 57, 218
O'Connor, Sandra Day, 214, 222
Office of Surface Mining Reclamation
    and Enforcement (OSMRE), 36
Ohio, 216
oil, 31, 53, 55, 82, 184
Oil Pollution Prevention Act (1990), 90
Oklahoma, 152
O'Leary, Rosemary, 5
Omnibus Public Lands Management
    Act, 15

O'Neil, Tip, 75
Oregon, 30, 155, 214
Owen, A. L., 11

Parson, Edward A., 172, 184
Pennsylvania, 31, 157
Pennsylvania v. Union Gas Company
    (1989), 157
People for the Ethical Treatment of
    Animals (PETA), 219
Percival, Robert V., 158
Persian Gulf War, 181
Philadelphia, PA, 156
Philippines, 173
Pilkey, Orrin, 175
Pinchot, Gifford, 8–9, 10, 11, 208
political partisanship
    in Congress, 62, 76, 77, 84–85,
        104, 220–21, 222
    and public opinion, 48–50, 51–52,
        53, 178
Pollution Prevention Act (1990), 31
Pombo, Richard W., 221
Powell, Lewis F., Jr., 214, 222
preservation, 8–9, 206
president
    and environmental policy, 3, 18–19,
        207–12
    roles of, 101–4, 107–11, 112–20,
        124
public opinion
    on climate change, 7, 17, 50–52,
        183–84, 217
    on economic growth, 48–49
    on environmental protection, 6, 18,
        48–50, 173–74, 217–18
    on governmental performance,
        54–55, 138–41
    and political partisanship, 48–50,
        51–52, 53, 178
    and Supreme Court, 162–63, 218
Purdy, Jedediah, 205, 206
Puyallup Tribe v. Department of Game
    of Washington (1968), 155

Reagan, Ronald
  and anti-environmental policy,
    13–14, 60–61, 108, 110, 208,
    210–11
  as chief executive, 113–14, 115,
    121, 160
  and the Court, 163, 222
  and economic development, 103,
    108
  and international agreements, 16,
    118, 120
Rehnquist, William, 222
Reid, Harry, 86
Republicans for Clean Air, 67
Republicans for Environmental
    Protection, 82
Republican Party
  in Congress, 14, 15, 61, 84–85,
    104, 108, 111, 213
  and public opinion, 48–50, 51–52,
    53
riders, 88–89
Ringquist, Evan, 5
Riverkeeper, 56
Roberts, John, 158, 214, 215
Roosevelt, Franklin D.
  as chief executive, 112, 113, 115
  and Congress, 102, 107
  as conservationist, 10, 120–21, 150,
    199–200, 206, 208
  environmental programs of, 11–12,
    99–100, 110
Roosevelt, Theodore, 4, 10, 11, 99,
    206, 208
Rosenau, James, 172
Rosenbaum, Walter, 8, 87, 90
Ruckelshaus, William, 137
rule making, 130–31, 135
Russia, 173

Safe Drinking Water Act, 3, 14, 212
Salt Lake City, UT, 217
Sansonetti, Thomas, 114

Scalia, Antonin, 158, 159–60, 167,
    214, 222
Schwarzenegger, Arnold, 28
scientific research, 7–8, 90–91, 93
sea level rise, 39–42, 175–76
Secretary of Interior v. California
    (1984), 154
Seelye, Katharine Q., 211
Sheeler, Kristina Horn, 4
Sierra Club, 10, 56, 62, 63, 206, 218,
    219
Sierra Club v. Morton (1972), 156
Silent Spring, 12, 23, 63, 150
Singapore, 119
Skocpol, Theda, 86
Slentz, Christina, 182–95
Soden, Dennis L., 107
Soil Conservation Service, 11, 100
Soil Erosion Service, 100
solidarity norms, 186–95
Sorenson, Theodore, 181
Sotomayor, Sonia, 214
South Carolina, 26
South Dakota, 215
South Korea, 119, 173
Southeast Florida Regional Climate
    Compact, 41–42
Soviet Union, 179
Sperling, Daniel, 216
Sporhase v. Nebraska ex re Douglas
    (1982), 152
States
  and the Court, 152, 154, 155, 156,
    157
  and federal programs, 5, 24, 25–27,
    30–33
  individual efforts in, 26, 27, 28–29,
    30, 31, 33–38, 39–42, 216–17
  See also individual states
Steel, Brent S., 107
Stevens, John Paul, 214, 222
Stossel, John, 56
strip mining. See mountaintop removal

Sundquist, James, 12
Sundstrom, Lisa McIntosh, 183
Superfund
 of 1980, 3, 13, 23–24, 90, 212
 of 1986, 3, 14, 88, 108
Supreme Court
 cases, 151–57, 167
 and environmental policy, 4–5, 13,
   19, 149–50
 and EPA, 152–53, 167, 213–14
 partisanship in, 221–22
 and public opinion, 162–63, 218
Surface Mining Control and
   Reclamation Act (1977), 34, 212
Sussman, Glen, 10, 50–51, 77, 114,
   184, 199
Swift, Art, 220
Switzerland, 119, 189

Tampa, FL, 176
Tatalovich, Raymond, 101
Taylor Grazing Act (1934), 12
Tennessee Valley Authority (TVA), 11,
   107, 163–65
Thomas, Clarence, 214
Tompkins, L. Forbes, 40
Toxic Substances Control Act (1976),
   161, 212
Train v. Campaign Clean Water (1975),
   152
Train v. City of New York (1975),
   151–52, 214
Truman, Harry S., 12, 102
Turkey, 188
TVA v. Hill (1978), 163–65

Udall, Stewart, 113
Underwood, Cecil, 35
unfunded mandates, 24, 27
Unfunded Mandates Reform Act
   (1995), 27
Union Electric Co. v. EPA (1976), 156
United Kingdom, 173, 179, 188

United Mine Workers of America, 34
United Nations Environment Program
   (UNEP), 177, 197, 201
United Nations Framework
   Convention on Climate Change
   (UNFCCC), 119, 185, 195
US Army Corps of Engineers, 12
US Forest Service, 208
US Navy, 153, 175, 181, 182, 214
U.S. v. Oregon (1935), 155, 214
U.S. v. Penn. Industrial Chemical
   Corporation (1973), 154
U.S. v. Riverside Bayview Homes
   (1985), 154–55
U.S. v. Ward (1980), 157
Utah, 14, 63, 216–17
Utility Air Regulatory Group v. EPA et.
   al. (2014), 153, 167

Van Putten, Mark, 66
Vermont, 154, 214, 215
Vietnam, 181
Vig, Norman, 121
Virginia, 175
Virginia Beach, VA, 176
voluntary programs, 24, 27, 30–33

Wallace, Henry, 100, 113
Wasby, Stephen, 163
Washington, 155
Water Quality Act (1987), 161
Water Resources Development Act
   (2007), 109
Waterkeeper Alliance, 56
Watt, James, 4, 114, 121, 126, 210
Weinberger v. Catholic Actions of
   Hawaii Peace (1981), 153
Weinberger v. Romero-Barcelo (1982), 153
Wenner, Lettie, 151
West Virginia, 33–38, 221
White House Office of Management
   and Budget (OMB), 130, 131
Whitman, Christine Todd, 4, 15

Wilderness Act (1964), 3, 63, 107
Wilderness Society, 62, 63, 219
Wilson, James Q., 91–92
Wilson, Woodrow, 73, 120
*Winter v. National Resources Defense
    Council* (2008), 153, 214
Wisconsin, 155, 215
Wisconsin Public Intervenor v. Mortier
    (1991), 155
Works Progress Administration, 100
World Trade Organization, 223

Wyly, Charles, 67–68
Wyly, Sam, 67–68
Wyoming, 215

Xi Jinping, 210

Yellowstone National Park, 65
Young, Oran, 176–77
Young, Rob, 175
Yulee v. State of Washington (1942),
    155